PAPERS OF THE

LABORATORY OF TREE-RING RESEARCH

NUMBER **4**

BRISTLECONE PINE IN THE
WHITE MOUNTAINS OF CALIFORNIA
Growth and Ring-Width Characteristics

Harold C. Fritts

The University of Arizona Press
Tucson, Arizona

The University of Arizona Press
www.uapress.arizona.edu

Printed in the United States of America
21 20 19 18 17 16   7 6 5 4 3 2

ISBN-13: 978-0-8165-0196-0 (paper)
ISBN-13: 978-0-8165-3521-7 (Century Collection paper)

L. C. No. 76-82614

♾ This paper meets the requirements of ANSI/NISO Z39.48-1992
(Permanence of Paper).

PREFACE

One of the principal research objectives at the Laboratory of Tree-Ring Research is the development of long, precisely dated tree-ring chronologies that may be used as indices of climatic variation. Ring widths from old trees growing on arid sites have been shown to be highly related to variations in annual precipitation and to a lesser extent temperature (Douglass 1928, Schulman 1956, Fritts 1965). A tree-ring chronology represents the average of dated ring-width patterns obtained from a number of trees. Each series of ring widths is transformed to relative index values by removing a growth curve. The index values for each year are pooled and averaged among all trees to obtain an average series or group chronology which, unlike the average of ring widths, is stationary through time (Fritts 1963, Matalas 1962).

During 1952, Edmund Schulman, a dendrochronologist at the Laboratory of Tree-Ring Research, was searching for old semiarid-site trees and discovered a subalpine-type limber pine (*Pinus flexilis* James) over 1,600 years old near Sun Valley, Idaho. He then began an intensive search for high-altitude species on dry sites. This search led him in 1953 to the bristlecone pine (*Pinus aristata* Engelem) on the dry upper slopes of the White Mountains in east-central California. These trees were older than the Sun Valley *Pinus flexilis* and exhibited highly variable (*sensitive*) and datable ring-width chronologies (Schulman 1956, 1958). Continued research in the area culminated in the discovery of an arid site (Methuselah Walk) where ages of some trees exceed 4,000 years. But Schulman's work in developing a reliable and long bristlecone pine chronology for the White Mountains was terminated by his sudden death in 1958.

His research was resumed by a cooperative project under the coordination of William G. McGinnies, at that time Director of the Laboratory of Tree-Ring Research. C. W. Ferguson continued to develop and extend Schulman's chronologies (Ferguson 1968, 1969), and I studied the relationships between ring growth and climate and the interactions with conditions of the site.

The investigation was supported by National Science Foundation grants GBI9949 and GP-2171. The field collection, ring dating, and data tabulations were in large part completed by John W. Cardis, graduate assistant on the project. C. W. Ferguson provided some materials and photographs as well as field and laboratory assistance. The author is indebted to W. G. McGinnies for his support, interest, and valuable counsel. Certain equipment and data were supplied by the U. S. Weather Bureau and by the U. S. Department of the Navy, China Lake, California. The White Mountain Research Station operated by the University of California, Berkeley, provided facilities and many services essential to completion of the project. I am also indebted to Harold A. Mooney and his associates as well as to Valmore C. LaMarche, Jr., Paul Campbell, Stanley Buol, William R. Boggess, and personnel of the Inyo National Forest, U. S. F. S., for their interest and help. The Numerical Analysis Laboratory, University of Arizona, provided free time on its digital computer.

HAROLD C. FRITTS

CONTENTS

CONTENTS

ILLUSTRATIONS

TABLES

PREVIOUS STUDIES

Bristlecone pine in the White Mountains of California is largely confined to cool dry sites at elevations between 9,500 and 11,600 feet (2,896 and 3,536 m). Tabulated data from the ten-year weather record at the White Mountain research station, located at 10,150 feet (3,094 m) elevation, have been presented by Wright and Mooney (1965). The mean January temperature is 20.4° F (-6.4° C) and the mean July temperature is 52.3° F (11.3° C). Approximately 10 inches (254 mm) of moisture falls each year as snow, and about 2.5 inches (64 mm) falls as rain.

The composition of a bristlecone pine stand in the White Mountains was described by Billings and Thompson (1957), whereas Wright and Mooney (1965) studied its distribution as related to substrate. Bristlecone pine occurs primarily on dolomitic limestone outcrops, though poorly developed stands share dominance with limber pine on two other substrates, a quartzitic sandstone and a granite. Wright and Mooney (1965) believe that the virtual restriction of bristlecone pine to dolomite substrates can be attributed to soil-water relations, although they acknowledge that the lack of competitors and tolerance for low nutrient availability may help account for its substrate orientation.

Mooney et al. (1962) and Wright and Mooney (1965) have demonstrated that the lighter colored dolomite has a 15 to 25% greater reflection than sandstone or granite. This greater reflection results in lower soil temperatures and lower evaporation; hence, a dolomitic soil remains consistently wetter than a sandstone soil when both are under the same climatic regime. The average rock cover upon the soil surface for sandstone is 84%, for dolomite 77%, and for granite 27%. No appreciable physical differences were noted between soils from north and south slopes. Mooney and his co-workers also found that the distribution of forest stands near the timberline is frequently related to the removal and deposition of snow by wind, as this influences the soil moisture supply.

From their measurements of photosynthesis and respiration, Wright and Mooney (1965) and Mooney et al. (1966) demonstrated that photosynthesis is sharply reduced only at soil moisture levels near 15 atmospheres tension and at light intensities below 3,000 foot-candles. Photosynthetic efficiency decreases as the trees grow older. Old trees show lower ratios of photosynthesis to respiration at all temperatures than do younger trees. These researchers estimated the volume of the green tops in old trees, which exhibited varying portions of living tissue in the main stem, and found that as the top dies the amount of live wood in the stem decreases. They concluded that the reduced photosynthetic efficiency of the leaves appears to be compensated for by trunk dieback. LaMarche (1969) states that cambial area reduction is favored by slow growth, wind damage, and soil erosion.

Schulze et al. (1967) have reported that during the onset of below-freezing temperatures in November, net photosynthesis in White Mountain pine declines. By mid-winter no net photosynthesis is observed, but respiration under warm conditions can be relatively high. As temperature increases in spring, and soils thaw, photosynthesis rates increase. Schulze and co-workers calculated that the trees they studied would require at least 117 hours of summer daylight to equalize the loss of energy due to respiration during the winter period.

Wright and Mooney (1965) have pointed out that along Methuselah Walk the environment is extremely harsh. They claim that the extreme aridity at this low elevation site, along with the slow growth rate of the trees, plays a part in the tree's longevity. They agree with Schulman's (1954) statement that for ancient trees longevity requires adversity, and also that the bristlecone pines in the White Mountains grow in a drier climate than do other stands of bristlecone pine to the east and south (Wright 1963, LaMarche 1969). Schulman (1956) was unable to locate trees in the 4,000-year age-class in areas east and south of the White Mountains.

More recently, Curry (1965) cut sections from a 4,900 year-old living tree in eastern Nevada, and LaMarche (1969) reports that very old bristlecone pines occur quite widely. However, trees with ages greater than 1,500 years are restricted to sites where edaphic and physiographic factors offset the effects of the regional precipitation. Old-age sites are frequently characterized by wide spacing of trees, compactness of tree crowns, sparcity of litter, and low density of accompanying ground-cover vegetation which provide a measure of protection from fire and competition.

Ferguson (1968) describes chronology building in the White Mountain bristlecone pine and discusses its application as a dating tool in radiocarbon analysis and archaeology. He has reported (1964) good crossdating among tree-ring chronologies of bristlecone pine, limber pine, and big sagebrush (*Artemisia tridentata* Nutt.) in the White Mountains. He believes that big sagebrush also reaches a maximum or near-maximum age in the White Mountains.

After analyzing the environment, several physiological processes, and some soil factors, Wright and Mooney (1965) concluded "that bristlecone pine is responding in gross terms primarily to a moisture gradient in the White Mountains. This would explain how its distribution runs counter to most of the subalpine herb and shrub components which seem to be controlled essentially by temperature gradients."

METHODS

The stands studied in this paper are mapped in Figure 1. They occur on dry and moist sites that range in elevation from 9,400 to 10,900 feet (2,865 to 3,322 m). Sixteen are on dolomite substrate, two (sites 2 and 3 in Fig. 1) on granite, and two (sites 5 and 18) on sandstone. Stands 2 and 18 are mainly limber pine, while 3 and 5 are largely bristlecone pine. A pinyon pine (*Pinus monophylla* Torr. and Frem.) stand (site 19) growing at the upper limits for the species is also included.

A saddle between two small peaks (sites 6-10, Fig. 1) was chosen for intensive growth measurements. Since the area is hidden from public view, it affords protection for instrumentation. The soil in the valley floor of the saddle is relatively deep and sufficiently free of rocks for moisture measurement (Fig. 2). Bristlecone pine representing a wide range of age-classes occupies both the valley floor and the adjacent north and south slopes.

ENVIRONMENTAL MEASUREMENTS

On April 24-27, 1962, instruments were installed on or near the floor of the saddle (hereafter referred to as *valley*), and records were obtained for the growing seasons of 1962 through 1964. A standard weather shelter and recording rain gauge were erected in a level area within a forest opening. A hygrothermograph was housed in the shelter and was checked weekly by comparing the record with readings from "max-min" thermometers. Average day temperatures were calculated from seven readings at two-hour intervals starting with 6 a.m. and ending with 6 p.m. Night temperatures were averages of the five two-hour readings, including 8 p.m. and 4 a.m.

An actinograph was exposed in a horizontal position on a rocky ridge at the head of the valley. Since this instrument was dismantled during the winter, slight differences between the mean and the maximum radiation values among the three years occurred because of differences in exposure, adjustment, and leveling of the instrument. The area above zero circumscribed by the radiation record was planimetered and converted to langleys per day. A cumulative anemometer was set up and read weekly. There were winds almost every day, but their velocities varied in the valley from one exposure to another, depending on wind direction. Therefore the weekly measurements of wind velocities at the one anemometer position were inadequate and are not included in the analyses discussed here.

Six, small, 100- to 200-year-old trees with datable ring-width sequences were chosen for intensive study. One was on the south-facing slope, one on the north-facing slope, and four were on the valley floor (Fig. 2).

A total of 24 Colman soil-moisture units were installed in undisturbed soil at two depths in each of three locations under the crowns of each tree on the valley floor. The upper unit was at a depth of 3 inches (8 cm) and the lower at 10 to 13 inches (25 to 33 cm) to measure moisture in the upper (A_1) and lower (AC) soil layers. During the growing season, the units were read periodically with a standard Colman soil-moisture meter. Gravimetric soil-moisture samples were taken at intervals throughout the first year to calibrate the Colman readings. Those readings were corrected for temperature, by use of the methods of Horton (1955), and plotted against the gravimetric determinations for 1962 to obtain a line of best fit. Soil-moisture percentages from Wright and Mooney (1965) for 1/3- and 15-atmospheres tension were found to approximate values of 87 and 7 on the corrected Colman scale (Horton 1955). A soil profile was described and roots were mapped, following the procedures of Fritts and Holowaychuk (1959), in a soil pit that was dug 3.3 feet (1 m) from a 36-inch (91 cm) dbh (diameter, breast-height) bristlecone pine.

Daily weather data for 1949 through 1964, collected at the Crooked Creek Laboratory of White Mountains research Station (site 2, Fig. 1), were used to calculate daily water balances (Thornthwaite and Mather 1955, Engelbrecht 1961). The calculations were based on a two-layer system with a total soil-moisture storage of 2 inches (51 mm), a soil-moisture surplus recession factor of 90%, and an upper layer with a maximum holding capacity of 0.5 inches (12.7 mm). Moisture is assumed to be removed from the upper layer at the calculated potential

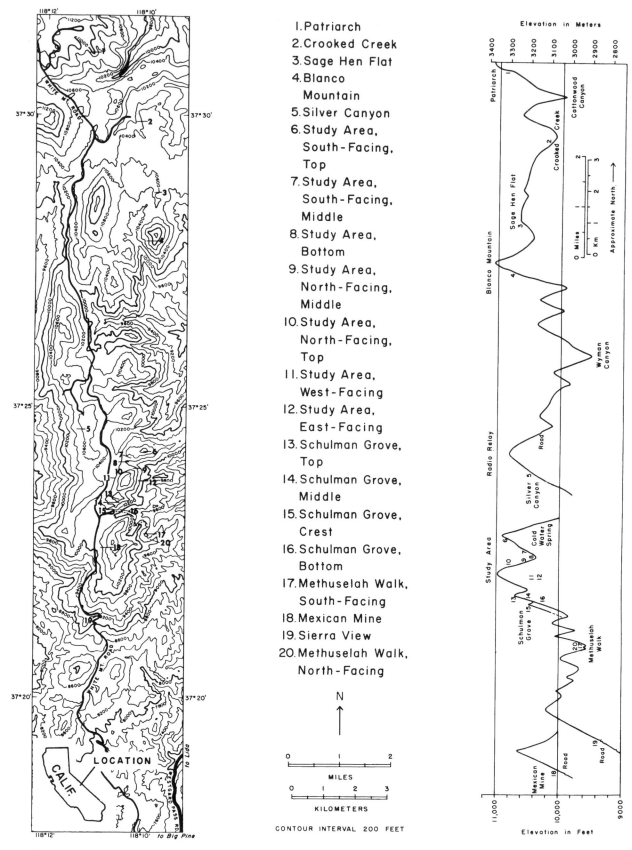

Fig. 1. Locations of sites sampled for tree-growth and tree-ring analysis in the White Mountains, California.

Fig. 2. The instrumented area (site 8 in Fig. 1) on the valley floor. In the foreground is a Colman soil-moisture meter. A dendrometer, which is being reset, and dendrograph, which is housed in the open shelter, are mounted on the bristlecone pine stem. The weather shelter is visible in the background. Note the vigorous full crowns of trees and the relatively dense stand indicating favorable growth conditions on this site.

evapotranspiration rate and from the lower layer in proportion to the percentage of available moisture. Although some details of this accounting system were considered not fully appropriate for the cold and arid climate of the White Mountains, the general scheme is thought to provide an approximation of the water budget.

GROWTH MEASUREMENTS

Six dendrographs (Fritts and Fritts 1955) equipped with battery-driven clocks were mounted approximately 60 inches (1.5 m) above the stem base of each tree (Table 1). Additional trees were selected for cambial sampling and for dendrometer measurements with the instrument described by Verner (1961). This type of dendrometer is essentially a simple lever constructed to retain the maximum reading since previous setting, so it was not necessary to make readings early in the morning in order to obtain the maximum size. A small screw was inserted into the outer bark surface at the contact point of the measuring arm for both dendrographs and dendrometers. This screw minimized fluctuations in size due to surface bark hydration. Mechanical difficulties interrupted and invalidated certain of these records in

TABLE 1

Site, Slope, and Size of Trees Measured by Dendrographs

Category	Site	% Slope	dbh (cm)	Height (m)
S F	South-facing slope	35	20.32	6.10
V 3	Valley floor	0	25.40	12.19
V 4	Valley floor	0	30.48	13.72
V 2	Valley floor	15	22.86	9.14
V 1	Valley floor	15	20.32	7.01
N F	North-facing slope	25	25.40	6.10

TABLE 2

Site, Slope, Average Size of Trees, and Number of Trees Measured by Dendrometers

Site	% Slope	dbh (cm)	Height (m)	Number of trees with valid records		
				1962	1963	1964
South-facing, young	25	18.62	6.71	3	0	2
South-facing, old	25	70.28	11.28	3	3	3
Valley, young	0	23.50	10.36	4	4	4
Valley, old	0	78.74	15.85	4	4	3
North-facing, young	35	25.40	7.01	3	3	2
North-facing, old	35	68.58	13.11	3	3	3

1963 and 1964. The trees measured by dendrometers represented both young and old age-classes. Each class was replicated four times on the valley floor and three times on each slope (Table 2).

During the growing season, samples for anatomic studies were collected periodically by means of a quarter-inch diameter leather punch from selected trees bearing dendrometers. The stem radii along cardinal directions were sampled and placed in F. P. A. (fixative solution Johansen 1940), and air was evacuated. Two to four trees were sampled in the valley and on each slope. In some cases, different trees were selected with each year to avoid excess wounding. At the end of the growing seasons for 1962 and 1963, selected dendrometers were moved to new positions and the rings under the contact points were harvested. At the end of the growing season for 1964 all radii measured by instruments were harvested.

Hand sections of anatomic samples collected throughout the growing season were obtained for at least two trees. The area near the cambium was stained with phloroglucin so that the lignified cells could be microscopically distinguished. Cells were observed and measured under high dry power with an ocular micrometer. Measurements were made of total xylem increment, number of lignified cells, and radial size of the last-formed row of lignified cells.

Four lateral branches on all trees bearing dendrographs were tagged for subsequent measurements of stem elongation and observations of phenological development. During the second season the leaders on four additional young bristlecone pines were also measured. Measurements were made from a reference mark on the stem to the end of the growing tip. Some difficulty in determining the actual stem tip was encountered between the time of bud opening and the spreading of young needles. During the latter two seasons special care was taken to distinguish between needle elongation and branch extension represented by increasing distance between the reference mark and the new apical bud. During 1963 additional phenological observations and cambial samples were obtained throughout the season from trees on sites 1, 3, 4, 5, 7, 9, and 16 (Fig. 1).

The lengths of needles and the lengths of twigs from trees on dry sites were observed to vary markedly from year to year (Fig. 3); the variations resembled those in the ring-width chronology. There-fore two primary lateral branches, one from the lower crown and the other from the upper crown, were collected from four trees in the Schulman Grove crest (site 15, Fig. 1). The stem segments delineated by terminal-bud-scale scars were dated among the sampled branches and against the rings shown in a cross-section through the segment. Measurements on each stem segment produced during the years 1949 through 1963 included length of each segment, total number of needles, average length of needles, and needle density (needles per unit length of segment).

TREE-RING ANALYSIS

In order to study changes in tree-ring characteristics that might be associated with site differences, 22 replicated samples of tree-ring sequences were obtained from twenty different sites (Fig. 1). Each sample was made up of measurements from at least 20 increment cores. (Two cores from each of 10 trees

Fig. 3. Bristlecone pine needles. Variation in needle length from one year to the next is common in arid sites. These needle lengths correlate with environmental variations from year to year and cross-date from tree to tree, as do the ring widths. (Courtesy of Keith Trexler, Chief Naturalist, Petrified Forest National Monument, Arizona).

represent a group.) For most samples, living bark completely encircled the tree stems (full bark). Trees were selected from relatively homogeneous areas. One core on each tree came from the side of the thinnest crown. Each increment core was surfaced and dated. Ring widths were measured starting with the 1860 ring through the last fully formed ring (1962 or 1963).

Ring widths were transformed to ring indices by the computer routine described by Fritts (1963). This procedure mathematically estimates the changes in width associated with increasing tree age. It then calculates the index by dividing each year's ring width by the estimate for that year. Indices for each core, unlike ring widths, are stationary with time and have a mean value of one (Matalas 1962). Therefore, they may be averaged to obtain chronologies for each tree and for each group.

The standardized series of indices for the two cores per tree were averaged to obtain a tree chronology. These were in turn averaged to obtain a group or site chronology. Three statistics were calculated from each group chronology. The standard deviation of the indices and the first order serial or autocorrelation were obtained in the usual manner. The former measures departures from the population mean and the latter measures the non-random association in the time series obtained from the covariance of each index with the index of the following ring. The mean sensitivity for each chronology is calculated as

$$\frac{1}{n-1} \sum_{i=1}^{n-1} \left| \frac{2(x_{i+1} - x_i)}{x_{i+1} + x_i} \right|$$

where x_i is a ring index and n the number of indices in each series.

The indices for each core were correlated with indices for all other cores in each group. The correlations among cores within each tree were averaged, as were the correlations among cores from different trees. These are referred to as the correlation for cores within trees and the correlation for cores among trees. All combinations of tree chronologies (average chronology of two cores) were correlated and the coefficients averaged to obtain the mean correlation for chronologies among trees.

Analyses of variance following Snedecor (1956) and Fritts (1963) were employed to obtain the estimated mean squares (EMS) due to chronology of the common group or site, due to differences among chronologies of trees, and due to differences among cores within trees. These three variance components were summed, and the percentage of variance attributed to each component was obtained. The mean error of individual indices is obtained from the component variances representing trees and cores.

The tree-ring chronology from each of the 22 groups, and the needle lengths, were correlated with the 15-year climatic record from the Crooked Creek Laboratory, using a stepwise multiple-regression analyses (Fritts 1962). The mean ring-width index or needle length for each year was associated with temperature, precipitation, and the calculated evapotranspiration deficit for the antecedent July through November, December through March, and the concurrent April-August. The latter period included the growing season, so it was subdivided into monthly and, in the case of August, half-monthly components in order to more precisely define the relationships.

COMMUNITY ANALYSIS

A square fifty-meter quadrat was established on each of seven sites. All stems one inch or more dbh were identified, number of stems per tree were recorded, and all diameters were measured. These data were converted to basal area and density values, calculated according to the method of Curtis (1956).

RESULTS

SOIL

The profile in dolomite-derived soils was described from road cuts as well as from the face of a soil pit dug in the gently sloping valley floor of the Study Area. Figure 4 shows the root numbers and sizes along with the soil horizons exposed on a two-foot wide segment of the vertical trench wall. The soil may be classified as a minimal calcisol (Harper 1957). Wright and Mooney (1965) report that the "dolomite-derived soils are alkaline, have a high content of calcium and magnesium, and are very low in available phosphorus."

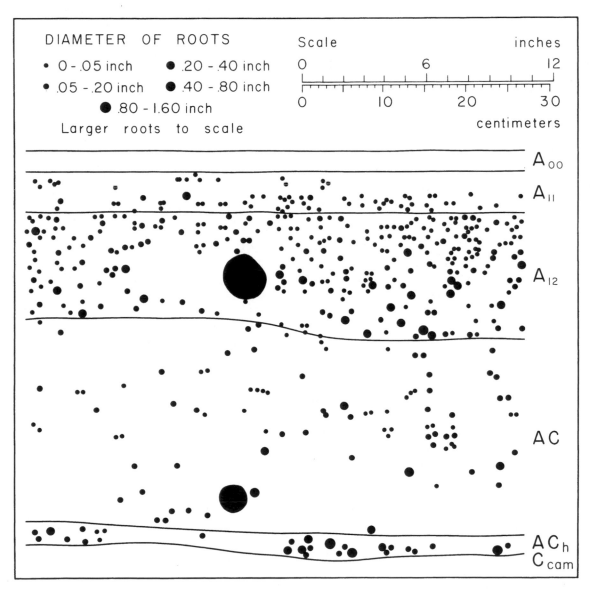

Fig. 4. Soil horizons and root distribution in the Study Area. Roots are from bristlecone pine and are most dense at depths of 5 to 6 inches and at the bottom of the profile immediately above the massive calcium layer. Rocks occupy a large portion of the soil mass.

The soil profile shown in Figure 4 is described in dry condition (Soil Survey Staff 1951) as follows:

A_{00} A thin layer of pine needles and twigs.

A_{11} 0 to 2 inches (0 to 5 cm), yellowish-brown (10 YR 5/4) loam with moderate very fine granular structure.

A_{12} 2 to 7 inches (5 to 18 cm), yellowish-brown (10 YR 5/4) very gravelly loam with moderate fine granular structure.

AC   7 to 17 inches (18 to 43 cm), yellowish-brown (10 YR 5/4) very gravelly loam with moderate fine granular structure and very pale brown (10 YR 8/3) coatings.

AC_h 17 to 18 inches (43 to 46 cm), dark reddish-brown (5 YR 3/2) very gravelly loam with moderate fine granular structure, humus rich layer (organic matter 15% dry weight, La Marche 1968).

C_{cam} 18 to 20 inches (46 to 51 cm), pinkish-white (7.5 YR 8/2) with pinkish-gray (7.5 YR 6/2) mottling, massive, accumulation of calcium carbonate.

C_{ca} 20 inches (51 cm)+, light yellowish-brown (10 YR 6/4) gravelly loam with weak fine granular structure.

As seen in Figure 4, the roots are most abundant in the A_{12} horizon between 2 and 7 inches (5 and 18 cm) and reach a secondary maximum in the organic AC_h horizon just above the C_{cam} carbonate layer.

This carbonate layer is extremely impervious, preventing deeper root penetration and water percolation. Roots are markedly less frequent or absent below the carbonate layer. Gravel and some cobbles, largely dolomite, are abundant at all depths and occupy more than 50% of the soil mass. LaMarche (1967, 1968) has described in detail the soil-texture variation.

The soil is low in clay, has a 20% total water storage capacity based on fines only (Wright and Mooney 1965), a rooting depth of 18 inches (46 cm), and a rock content exceeding 50%. Since the soil is underlain by an impervious C_{cam} horizon it is concluded that the maximum soil moisture retention is approximately 2 inches (5 cm). In order for this type of profile to develop, a large part of the moisture that falls as snow must evaporate or run off. Two or three inches of moisture are retained in the soil during the early spring period, sometimes producing saturated conditions and a perched water table in the lower AC and AC_h horizons. Such conditions of saturation at low levels were observed during installation of the soil moisture units on April 24-27, 1962, and were measured gravimetrically throughout the following June.

VEGETATION

The quadrat data are summarized in Tables 3 and 4. They include sites from which tree-ring samples were obtained. These results are in general agreement with the stand figures of Wright and Mooney (1965).

Both studies document the fact that bristlecone pine occurs in almost pure stands and reaches its greatest density on north-facing dolomite slopes. On the dryer south-facing dolomite slopes, trees have somewhat

TABLE 3

Stand Characteristics of Four Selected Bristlecone Pine Communities Growing on Dolomite in the White Mountains

Sample	Basal area per hectare (sq. m.)			Number of trees per hectare			Basal area per tree (sq. m.)			Percent with multiple stems		
	BCP*	LMP†	Total	BCP	LMP	Total	BCP	LMP	Total	BCP	LMP	Total
South-facing slopes	23.96	0.95	24.91	95	3	98	0.25	0.32	0.25	38	67	39
North-facing slope	47.46	0.00	47.46	132	0	132	0.36	0.00	0.36	24	00	24
Sandstone	15.77	6.62	22.40	44	24	68	0.36	0.28	0.33	36	17	29
Granite	8.85	13.51	22.36	20	44	64	0.44	0.31	0.35	40	9	19

*BCP — Bristlecone pine.
†LMP — Limber pine.

TABLE 4

Percentage of Trees per Diameter Class in Four Selected Bristlecone Pine Communities Growing on Dolomite *

Sample	Diameter at breast height (cm.)							Total Individuals
	2.5 - 12.5	12.5 - 25.0	25.0 - 50.0	50.0 - 75.0	75.0 - 100.0	100.0 - 125.0	125.0 - 150.0	
South-facing slopes	8	20	28	27	13	2	2	98
North-facing slope	3	9	24	33	18	9	3	33
Sandstone	18	0	23	29	18	12	0	16
Granite	25	6	0	38	25	6	0	17

*Multiple stem diameters were converted to cross-sectional area, totaled for the tree, and each was classified according to equivalent area of single stem trees.

smaller basal areas, and multiple stems are more likely to occur.

Limber pine is co-dominant on sandstone but becomes dominant on granite. These non-dolomitic substrates support a more open stand in which bristlecone pine is sparsely distributed and more influenced by inter-specific competition and local conditions of the site.

The measurements of stem size (Table 4) show a remarkably uniform distribution of stem diameters up to 100 cm (39 inches), which approximates an age of 1,000 years. Decreased densities at diameters greater than 100 cm indicate increasing mortality perhaps associated with the beginning of crown and bark dieback in the aging trees.

The quadrat data show that 23% of all sampled trees have two main stems, 4% have three, and 7%

have four or more. In addition to multiple stems, portions of the cambium may die in many old trees, leaving only a thin strip of bark connecting the live roots and crown. These so-called strip-bark trees are especially common on exposed sites and are evidence of extreme site adversity (Schulman 1954, Wright and Mooney 1965, La Marche 1969).

Comparisons with Wright and Mooney's cover data show that our samples are representative of the range for bristlecone pine stands on dolomite in the White Mountains. However, densities in our samples are higher for sandstone and granite sites. We deliberately chose abnormally dense sandstone and granite stands in order to obtain a sufficient replication for the analysis of tree rings. Wright and Mooney sampled average substrates, which included large areas with few standing trees.

CLIMATE

Although the White Mountains form a rather large and high mountain mass, they lie in the rain shadow of the Sierra Nevada. The annual precipitation is low, and the total amount is highly variable from year to year. The climatic record for 1949 to 1964 shows that precipitation for the period December through March varied from 1.35 inches (34 mm) in 1961 to 9.83 inches (250 mm) in 1949. A month with no precipitation may occur at any time. The highest recorded monthly precipitation during June through September is generally lighter and may fall as either rain or snow.

Temperatures average below freezing from November through April. The average maximum temperature is near freezing for January and February, although maximum temperatures above 50° F (10°

C) have been recorded during all months of the year. Minimum temperatures on the average are above freezing from June through September, but frost and freezing temperatures may occur. Maximum temperatures rarely exceed 76° F (24° C), and minimum temperatures during the winter are seldom lower than -20° F (-29° C). The climate at the Crooked Creek Laboratory of the White Mountain Research Station (site 2, Fig. 1) is described in more detail by Wright and Mooney (1965).

Figure 5 includes plots of average temperatures, total precipitation, calculated average available soil moisture, and calculated total evapotranspiration deficit for May through August of 1962, 1963, and 1964, with plots of station averages for 1949 to 1964. All values were derived from the Crooked

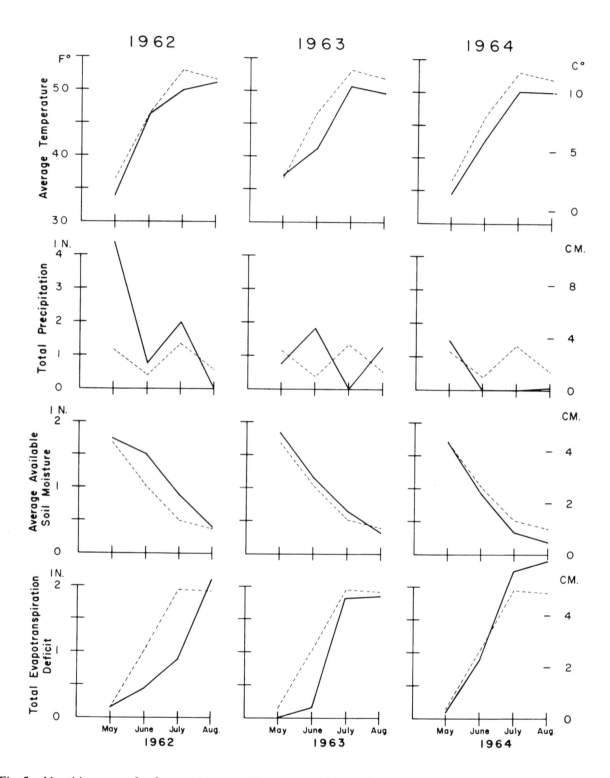

Fig. 5. Monthly means for four environmental parameters during the three study years (*heavy line*) plotted with the 15-year means (*dashed line*) for the months May through August. All data are based on or calculated from records at the Crooked Creek Field Station.

Creek record. It may be seen from these data that mean monthly temperatures for the summer season were less than the station averages during all three years of this study. Precipitation and soil moisture were well above average in 1962 and below average in 1964. Evapotranspiration deficits were considerably below average in 1962 and 1963 and exceeded the average only during August of 1962 and during July and August of 1964.

It is apparent from the figure that even though precipitation varied markedly between the three growing seasons, low temperatures prevented rapid evapotranspiration, so that drought conditions became limiting only during the latter part of 1964. Evapotranspiration deficit calculations showed that this drought was considerably less extreme and occurred later in the growing season than two other droughts which occurred during 1950 and 1960 and which were accompanied by very narrow growth rings.

Figure 6 shows the average day and night temperatures, total radiation, daily precipitation, and average soil moisture readings at two depths measured at the Study Area (site 8, Fig. 1) for the three growing seasons. Daily evapotranspiration deficits shown in the figure were calculated from the Crooked Creek weather record which sometimes differed from our Study Area data due to the local nature of summer convectional storms. Two snowstorms, one in mid-June and the other in mid-July, resulted in abundant soil moisture throughout 1962. As a result of May precipitation and low June temperatures in 1963, soil moisture remained above field capacity in June and did not become critically low until late in the season. In 1964 soil moisture was low by mid-June, and by July it reached wilting percentage at the 3- and 10-inch depths (76 & 254 mm). Rain on August 4 and 5 replenished soil moisture in the Study Area but was very light at the Crooked Creek weather station, so it is not apparent in the calculations of evapotranspiration deficit.

RADIAL GROWTH AT THE INSTRUMENTED AREA

Dendrographs, dendrometers, and cambial samples provided measurements of changes in the water balance and the radial growth of the trees. Dendrographs recorded stem expansion at night or during cloudy and moist weather when water absorption exceeds the water lost from the tree (MacDougal 1936, Fritts 1958). Stem shrinkage was common in the daytime whenever transpiration exceeded the absorption of water by the tree roots. Ordinarily the maximum stem size occurs around 6 a.m. and the minimum between 3 and 6 p.m. The relative changes in stem radius recorded by each dendrograph are plotted (Fig. 7) in a fashion to represent the individual maximum and minimum size recorded each day. The letters in Figure 7 designate various phenological stages of the trees.

Early in the season the cambium is dormant so that expansion and contraction in the stem are approximately equal throughout the 24-hour day. Dramatic shrinkage, apparently due to freezing in the stem, occurs whenever night temperatures drop as low as 20° F (-7° C). Stems expand considerably during precipitation periods, such as occurred June 15, 1962; but with the return of clear weather, stems shrink to their original size. As soon as swelling is observed in the terminal buds, cambial activity commences and stem expansion at night exceeds shrinkage during the day. Diurnal fluctuations in the stem size reach their greatest magnitude when soil moisture is low and evapotranspiration deficits are high (Figs. 5 & 6).

The growth rate or net increase in size is most rapid at the time of bud elongation and begins to decrease as the new needles emerge. By early August, the time of pollination, little further increase in stem size may be recorded. However, in 1964, water stress within the tree delayed late-season cell expansion. With the advent of precipitation in mid-August, favorable water relations were reestablished and rapid expansion of the last formed cells must have occurred.

Dendrographs with battery-powered clocks recorded stem-sizes for the winter of 1962-1963. Periods of extreme stem shrinkage accompanied cold weather. This shrinkage apparently results from ice formation in the xylem and dehydration of the living tissues in the stem. Slight expansion occurred when ice turned into water at the beginning of each thawing cycle, but in a few days an osmotic equilibrium was established and each radius returned to its original size prior to the freeze.

Plots of average maxima and minima recorded by the four dendrographs on valley trees (V1-V4, Fig. 7) are shown in Figure 8, along with dendrometer and cambial measurements for 1962. All three growth measurements exhibit good agreement. Several of the

14

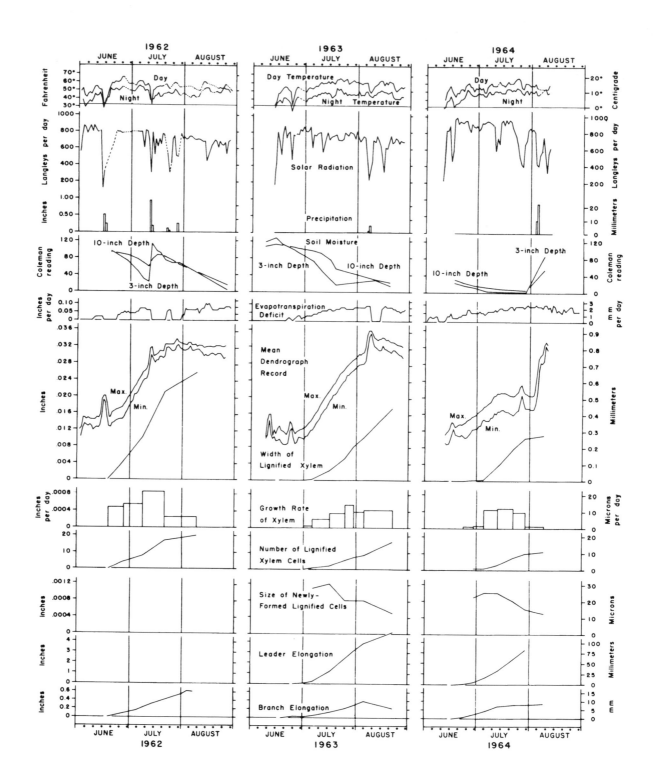

Fig. 6. Daily march of environmental and growth parameters measured for the three growing seasons. Evapotranspiration deficits were calculated from the Crooked Creek records; all other measurements were made at the Study Area (*bottom*) site.

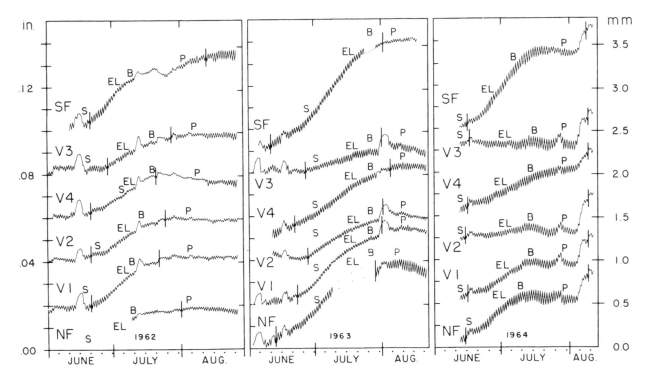

Fig. 7. The daily maximum and minimum stem sizes as measured during the three growing seasons by dendrographs mounted upon six young bristlecone pine. Tree locations: SF—South-facing site; VI, V2, V3, V4—Valley floor; and NF—North-facing site. Phenological stages: S—buds swelling; EL—buds elongating; B—buds opening, needles starting to emerge; and P—pollen shedding. Vertical lines designate time of 5 and 95% of total stem enlargement.

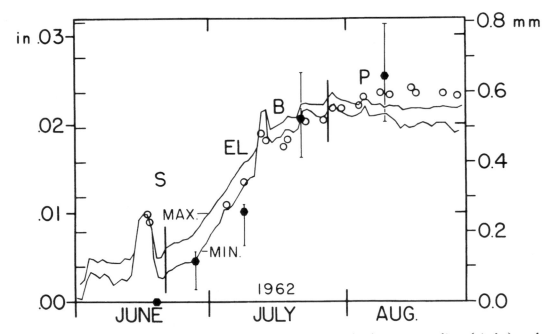

Fig. 8. Average results for dendrograph measurements *(solid lines)*, dendrometer readings *(circles)*, and cambial samples *(hexagons)* during 1962 for trees on the valley floor. Heavy vertical lines designate time of 5 and 95% growth as measured by dendrographs. Light vertical lines show range of cambial measurements.

dendrometer measurements and one of the two cambial samples exhibited slightly more growth early in August than that recorded by the dendrographs. Growth stimulation due to wounding from the cambial punch and from the inserting of the dendrometer screws is the most probable explanation for these differences (Fritts 1958). Additional checks on the three measurements of growth during the 1963 and 1964 seasons revealed similar results.

The sections under the nine dendrometers that were relocated after the 1962 growing season provide further comparison of measuring techniques (Fig. 9). Ring widths and stem increments measured by the dendrometers exhibit a 1:1 relationship approximating the x = y line in Figure 9. The two dendrometers that gave large departures were subject to large reading and reset errors. A further check made from samples taken at the end of the 1963 season produced essentially the same results. The dendrometers used in this study were designed to measure the maximum stem size between reset times. Thus the minimum size recorded by these instruments is only by chance the actual minimum occurring during the growing season. This causes dendrometer measurements of total annual growth to often be less than those obtained from dendrographs or those made directly from anatomic samples (Fig. 8). No significant contribution from phloem and cork production was noted in this study.

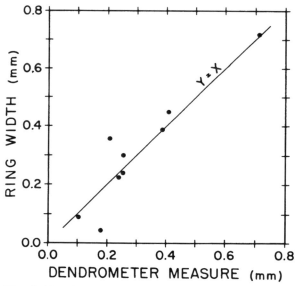

Fig. 9. Total growth as measured by several dendrometers plotted against the width of the xylem increment formed under the point of measurement by the end of the 1962 growing season. The line y = x approximates the relationship.

The dendrograph measurements on the valley trees were pooled for each year, and their means are plotted in Figure 6. Also shown are data for the accumulated width of the lignified xylem and its rate of enlargement during each sampled interval, the accumulated number of lignified cells, the size of the outermost newly formed lignified cells, elongation of terminal branches (leader), and elongation of lateral branches of the selected trees.

Growth during the three seasons is dependent on both temperature and moisture regimes. Cambial activity is initiated late in June or early July after freezing night temperatures cease to occur. The coldest June was in 1963. Growth initiation during that year was delayed by ten days to two weeks, but it continued later in the season; so during all three years the length of the growing period was approximately the same.

The maximum rates of growth, as measured by both dendrographs and anatomic samples, were higher during the moist and cool seasons of 1962 and 1963 than during the drier season of 1964. However, both the 1962 and 1964 seasons were characterized by low levels of soil moisture in early July and by a marked reduction in the number of xylem cells produced by the end of July. The mid-season storms of 1962, which replenished soil moisture, had little effect on either the length of the growing season or the rate of growth in these trees. Some exceptions were noted where growth was resumed, but only in the younger trees.

While growth in the leaders continues longer than in lateral branches, there is a clear relationship between the period of shoot activity and cambial growth during all three years. These results suggest that the growing period for this species is not necessarily shortened by depletion of soil moisture reserves or by day-length changes, but is frequently controlled by some phenological or internal factor that is closely timed with growth initiation. However, the number and sizes of cells produced under different moisture regimes are not the same.

The sizes of xylem cells formed along the six radii monitored by the dendrographs were measured from sections obtained when the instruments were removed in 1964. The cell widths along a radial direction were measured for five rows of cells in each ring. Measurements from the five cell rows in each layer were averaged, and then values of the three adjacent cells were averaged to obtain a series of running means. These mean cell sizes are plotted in Figure 10.

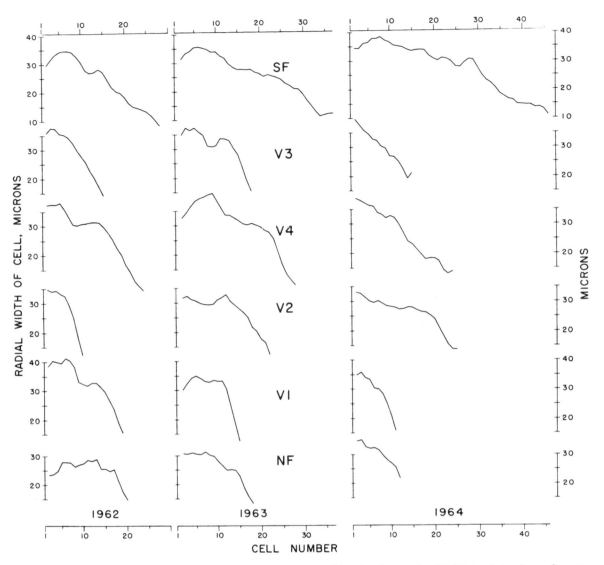

Fig. 10. Mean width of the cells formed in the rings monitored by dendrographs. Width is plotted as a function of cell number from the first-formed to last-formed for the season. Five rows of cells were measured in each ring; plots represent overlapping means for sets of three adjacent cells in each series.

A cell-size gradient is well marked in all rings. No clear boundary or sudden change in size between earlywood and latewood within the same ring can be discerned, but boundaries between adjacent rings are well defined. The earlywood generally contains more large cells during the moist years of 1962 and 1963 than in 1964, although an exception was seen in the rings from the north-facing slope, where the first-formed cells are smaller, especially in 1962.

It is possible that during the cooler years low temperatures were limiting to cell size growth in trees on the north-facing slope. Four of the six radii for 1962 exhibit a slight increase in cell size in the mid-portion of the ring. This change may be cor-

related with replenishment of soil moisture in July. A similar mid-season pattern that can be observed in two radii during 1963 and one in 1964 cannot be associated with an environmental change. In the 1964 rings, the last row of lignified cells for V3 and V4 exhibited a slight increase in size, and the last measured lignified cells were not as small as in the preceding season. This suggests that in 1964 the rain of August 5 allowed further expansion of the last two rows of xylem cells in two trees, and at the time of harvesting (August 12) lignification of the last formed row of cells was not complete.

It is clear from the graphs in Figure 10 that the sequences from wide earlywood cells to narrow latewood cells is highly variable among trees, especially during moist years, and is not a dependable criterion for evaluating variable climatic regimes. The cell-size changes measured within a ring were not of sufficient magnitude to be considered typical of "false" rings. In fact, *false* rings or intra-annual latewood bands which may be confused with true

boundaries between annual rings in bristlecone pine from the White Mountains have been observed by LaMarche only in two trees (personal communication). Ferguson (1968) reports that he has "found no more than three or four occurrences of even incipient multiple growth layers."

RELATION OF GROWTH TO SLOPE ASPECT AND TREE AGE AT THE INSTRUMENTED AREA

The growth measurements from dendrometers are summarized in Table 5 according to slope aspect and tree age (see also Table 2). The measurements obtained from anatomic sections of the radii monitored by both dendrometers and dendrographs are presented in a similar fashion in Table 6. Approximately two-thirds of the dendrometers had been moved to new radii during the three-year period, while the anatomic measurements were obtained from the same radii determined by the location of the instruments during the third year. As explained earlier, the annual growth increment from dendrometers sometimes underestimates the direct measurement of xylem width, as shown in Table 6.

Although the averages from the total sample for three years of dendrometric measurements show a growing season of 45 days starting on June 26 and ending on August 9 (Table 5), there are differences in the means classified according to age. The younger trees appear to initiate growth four to seven days earlier and cease growth as much as 13 days later than the older trees. The length of the growing season ranged from 47 to 56 days for the younger trees and 35 to 43 days for the older trees. The greatest differences among trees in the growing season occurred during 1962 when precipitation replenished soil moisture in July and apparently prolonged growth only in the most vigorous and youthful trees. Also,

TABLE 5

Summary of Average Dendrometer Measurements for the Three Growing Seasons
Classified According to Site and Tree Age*

| | Growing Season | | | | | | | | | | | |
| | 1962 | | | | 1963 | | | | 1964 | | | |
Sample	Beginning	End	Season length	Growth (mm)	Beginning	End	Season length	Growth (mm)	Beginning	End	Season length	Growth (mm)
South-facing slope												
Young	6/25	8/20	57	0.937	6/14	8/11	59	..	6/24	8/9	47	0.538
Old	6/25	8/9	46	0.295	7/4	8/11	39	0.461	6/30	8/9	41	0.345
Average	6/25	8/14	51	0.616	6/24	8/11	49	..	6/27	8/9	44	0.442
Valley bottom												
Young	6/25	8/17	54	0.462	6/19	8/11	54	0.536	6/24	8/9	47	0.423
Old	7/6	7/29	24	0.127	7/4	8/11	39	0.334	6/24	8/9	47	0.401
Average	7/1	8/8	39	0.295	6/27	8/11	46	0.435	6/24	8/9	47	0.412
North-facing slope												
Young	6/20	8/9	51	0.414	6/19	8/11	54	0.586	6/24	8/9	47	0.325
Old	6/25	7/29	35	0.231	7/4	8/11	39	0.411	6/30	8/9	41	0.344
Average	6/22	8/3	43	0.322	6/27	8/11	46	0.499	6/27	8/9	44	0.335
Total sample												
Young	6/23	8/15	54	0.605	6/17	8/11	56	0.561	6/24	8/9	47	0.429
Old	6/29	8/2	35	0.218	7/4	8/11	39	0.402	6/28	8/9	43	0.363
Average	6/26	8/8	44	0.412	6/26	8/11	47	0.483	6/26	8/9	45	0.396

*The numbers and location of dendrometers are shown in Table 2.

TABLE 6

Summary of Average Ring Widths and Number of Cells in the Growth Layers,
Monitoried by Dendrographs and Dendrometers, Classified According to Site, Tree Age, and Year

| Sample | Growing Season | | | | | | | |
| | 1962 | | 1963 | | 1964 | | Average | |
	Ring width (mm)	No. of cells	Ring width (mm)	No. of cells	Ring width (mm)	No. of cells	Ring width (mm)	No. of cells
South-facing slope								
Young	1.029	32	1.044	36	0.790	29	0.954	32
Old	0.376	11	0.566	18	0.384	12	0.442	14
Average	0.702	22	0.805	27	0.587	21	0.698	23
Valley bottom								
Young	0.549	18	0.604	19	0.546	19	0.566	19
Old	0.254	8	0.678	24	0.472	17	0.468	16
Average	0.402	13	0.641	22	0.509	18	0.517	18
North-facing slope								
Young	0.444	16	0.564	19	0.470	19	0.493	18
Old	0.285	10	0.569	20	0.490	17	0.448	16
Average	0.364	13	0.567	20	0.480	18	0.470	17
Total sample								
Young	0.674	22	0.737	25	0.602	22	0.671	23
Old	0.305	10	0.604	21	0.449	15	0.453	15
Average	0.489	15	0.671	23	0.525	19	0.562	19

the mean growing season in 1962 continued 11 days later in trees on the south-facing slope than in trees on the north-facing slope. Rain in early August of 1964 rehydrated all trees, causing an apparent but short increase in growth rate, which was followed by simultaneous cessation in growth for all sampled trees (Figs. 6 & 7).

Average ring widths and corresponding cell numbers are greatest for the moist 1963 growing season (Table 6). The younger trees have wider rings than do the older trees, but there is less difference in the relative ring widths from year to year. Comparisons of ring widths among all three sites show that the widest rings were formed in younger, south-facing slope trees, while the narrowest rings were produced in the old, south-facing slope trees. Differences observed among age-classes were least marked on the north-facing slope. The 1962 growing period for vigorous young trees bridged the dry period in early July so that a wide ring was formed. In the cooler seasons of 1963 and 1964 the young trees on the south-facing slope started growth earlier and grew more rapidly (Fig. 7) than the other trees. During the dry 1964 season, moisture became limiting everywhere, and ring widths were similar among age-classes and sites.

PHENOLOGICAL DEVELOPMENT AND GROWTH ON OTHER SITES DURING 1963

Phenological development was observed on four comparable trees per site in eight areas (sites 1, 3, 4, 5, 7, 8, 9, & 16, Fig. 1). They ranged in elevation from 10,000 to 11,000 feet (3,050 to 3,350 m) and included dolomite, sandstone, and granite substrates, and north, south, and west exposures.

Initial bud swelling began earliest (June 20) in the low elevation and granite sites (16 and 3, respectively) and about 10 to 12 days later in the highest and in the north-facing sites (1 & 9, respectively). The cambial zones began to enlarge at approximately the same times. The first new cells were observed about two days later.

Bud elongation was initiated approximately July 8-10 in all but the higher elevation sites, where elongation was delayed about eight days. This stage accompanied lignification of the first row of tracheids.

Needles emerged from under the bud scales approximately July 28-30, except at the higher elevations, where equivalent development was delayed about eight days.

Pollination started August 6-8 on all but the high elevation sites, where it followed in approximately eight days. When this stage is attained, needles are almost fully elongated but they remain appressed to one another in the fascicle; terminal growth is complete and most all cells are lignified.

On August 17 all cells were formed and lignification of the last-formed latewood was in progress except on the highest elevation site.

Apical growth at the lowest elevation, sandstone, and granite sites was approximately double the apical growth at other sites. The least apical growth occurred on the north-facing slope and the two highest elevation sites. Growth of lateral branches was greatest on granite and least on the north-facing slope.

The most rapid radial growth occurred on the granite, and the least rapid radial growth occurred on the lower elevation, south-facing slopes. The growth rate at the highest elevation site was only slightly greater than in the low elevation trees, but the growing season was later and perhaps somewhat shorter.

TREE-RING CHARACTERISTICS

Statistical characteristics of the tree-ring samples are presented in Table 7. The statistical characteristics are similar to those described by Fritts (1965), Fritts et al. (1965a, b). Mean ring width, mean stem diameter at breast height (dbh), and the percentage of rings missing (locally absent or partial rings) are averages from the ten trees used in each sample. The standard deviation of the indices estimates the variation for each group chronology as measured from the sample mean, and mean sensitivity is a relative measure of first differences between adjacent rings. First-order serial correlation (r) measures non-random association of indices from year to year. Correlation for cores within trees (r cores within trees) measures the similarity among cores within the same tree. The correlation of cores between trees measures the similarity among cores from different trees, while the correlation among trees measures the similarity in tree chronologies for each group. The correlations among the 19 bristlecone pine group chronologies were obtained, and the mean correlation of each one with the other 18 is shown in Table 7 (as r with other groups).

The estimated mean square component of each chronology labeled Chronology EMS in Table 7 is analogous to the chronology standard deviation, and its percentage of total variance is given in the fourth column from the right. The remaining variance component percentages represent differences among the chronologies of individual trees and among the individual cores within the trees for each group. The error of individual indices in each group chronology is shown on the right.

Variation Due to Tree Age

The data for site 8, the first two lines in Table 7, provide a comparison between younger full-bark trees (150 to 300 years old) and older full-bark trees (averaging approximately 600 years) which are located in the valley of the Study Area (Fig. 2). The two chronologies are shown in Figure 11. As would be expected, the younger trees have generally wider rings, somewhat fewer missing rings, less chronology variation, less serial correlation, and lower correlation among cores and among trees, as well as lower correlation with other sampled sites. For the younger trees the chronology EMS is lower and represents only 44% of the total variance. Only 13% of the remaining variance is due to differences in tree chronologies. The remaining 41% is due to the large differences among cores within trees which can be attributed to differences in growth rates and growing seasons which were measured in young trees (Tables 5 & 6).

Variation Due to Slope Aspect Within the Instrumented Valley

Sites 6, 7, 9, and 10 (lines 3-6, Table 7) represent samples from the upper and mid south-southwest-facing and upper and mid north-northeast-facing

TABLE 7

Statistical Measurements of Chronology Characteristics for Twenty-two Replicated Tree-Ring Samples
from Twenty Sites in the White Mountains of California

Site No.	Name	Mean ring width (mm)	Mean dbh (mm)	Rings missing (%)	Standard Deviation	Mean sensitivity	First order serial (r)	r Cores within trees	r Cores between trees	r among trees	r with other groups	Chronology EMS	% Variance Chronology	Trees	Cores	Error of individual indices
8	Study area, young	0.56	246	0.82	0.228	0.236	0.23	0.60	0.46	0.58	0.83	0.049	44	13	41	0.061
8	Study area, old	0.45	622	0.86	0.241	0.245	0.28	0.66	0.52	0.62	0.87	0.055	50	19	31	0.062
6	Study area, south-facing, top	0.43	676	1.55	0.278	0.313	0.20	0.80	0.68	0.75	0.88	0.075	66	15	19	0.053
7	Study area, south-facing, middle	0.38	366	1.00	0.241	0.263	0.13	0.77	0.63	0.71	0.87	0.056	61	14	25	0.049
9	Study area, north-facing, middle	0.38	533	0.75	0.202	0.205	0.22	0.59	0.46	0.58	0.78	0.038	45	13	43	0.054
10	Study area, north-facing, top	0.31	361	3.35	0.280	0.328	0.16	0.72	0.58	0.66	0.85	0.074	56	20	25	0.062
11	Study area, west-facing	0.80	386	0.35	0.209	0.199	0.33	0.73	0.59	0.68	0.86	0.041	58	18	25	0.045
12	Study area, east-facing	0.38	356	0.40	0.243	0.241	0.28	0.72	0.58	0.68	0.86	0.056	58	15	28	0.049
15	Schulman Grove, crest	0.26	··	3.75	0.369	0.426	0.21	0.85	0.68	0.74	0.84	0.131	67	17	16	0.070
13	Schulman Grove, top	0.53	556	0.15	0.232	0.221	0.30	0.75	0.57	0.65	0.81	0.051	55	19	27	0.051
14	Schulman Grove, middle	0.53	434	0.80	0.254	0.233	0.35	0.65	0.53	0.64	0.85	0.061	48	9	43	0.052
16	Schulman Grove, bottom	0.49	538	1.10	0.260	0.282	0.11	0.72	0.57	0.66	0.85	0.064	55	16	29	0.060
1	Patriarch	0.33	597	0.40	0.243	0.244	0.33	0.70	0.55	0.65	0.76	0.056	54	15	31	0.061
4	Blanco	0.34	726	0.30	0.228	0.227	0.29	0.71	0.55	0.64	0.82	0.049	53	17	30	0.056
17	Methuselah Walk, south-facing	0.34	234	1.78	0.295	0.345	0.17	0.79	0.63	0.70	0.82	0.083	61	17	21	0.070
3	Granite	0.56	800	0.30	0.290	0.224	0.59	0.77	0.61	0.69	0.75	0.080	61	15	24	0.060
5	Sandstone	0.51	744	1.00	0.226	0.217	0.34	0.67	0.48	0.56	0.82	0.048	48	23	29	0.062
20	Methuselah Walk, complacent	0.29	··	1.44	0.241	0.242	0.11	0.53	0.41	0.53	0.80	0.052	38	14	49	0.076
20	Methuselah Walk, sensitive	0.23	··	3.42	0.288	0.321	0.11	0.62	0.53	0.65	0.76	0.078	50	11	39	0.084
2	Limber pine	0.47	465	0.25	0.216	0.171	0.53	0.72	0.52	0.61	··	0.044	52	19	29	0.054
18	Limber pine	0.44	521	0.70	0.243	0.235	0.26	0.68	0.53	0.63	··	0.056	50	16	34	0.051
19	Pinyon pine	0.66	295	2.95	0.367	0.430	0.34	0.72	0.50	0.57	··	0.124	46	22	32	0.106

slopes in the same area as site 8 (Fig. 11). The sites on the upper slopes are both windswept and relatively rocky; the lower sites especially on the north-facing slope and valley bottom are more protected. The high percentages of missing rings in the chronologies from the upper slopes, high standard deviation, and high mean sensitivity (Table 7) indicate that climate has frequently limited growth of these trees (Fritts et al. 1965b). The chronologies from the south-facing slope correlate best with one another and with all other bristlecone pine groups. While the chronology variance is similar in both of the upper samples, it represents 10% more of the total group variance in the upper south-facing than in the upper north-facing slope trees. Characteristics of the south-facing mid-slope stand suggest intermediate aridity, while those from the north-facing mid-slope indicate more moist

conditions than even the valley bottom (site 8). These differences are reflected by decreased percentages of missing rings, reduced standard deviations, mean sensitivities, correlations among cores, trees, and groups, and by increased variance for cores within trees.

Sites 11 and 12 (lines 7 and 8, Table 7) represent samples from the west- and east-facing slopes near the Study Area (Figs. 1 and 12). The west-facing stand actually faces west-northwest and is on a steep slope (40°) which rises out of a treeless sagebrush flat. The wide rings, low percentage of missing rings, and low chronology variability of the west-facing trees indicate abundant moisture in this site. The blowing snow from the sagebrush flats to the west forms drifts on this site, and ample melt water is present every spring. In the west-facing as well as the north-facing

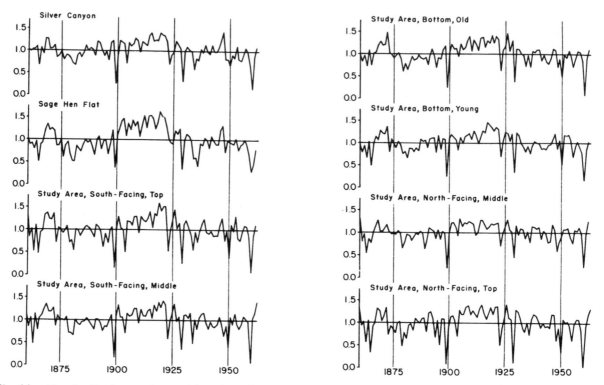

Fig. 11. Standardized tree-ring width indices for the chronologies of eight bristlecone pine stands in the White Mountains.

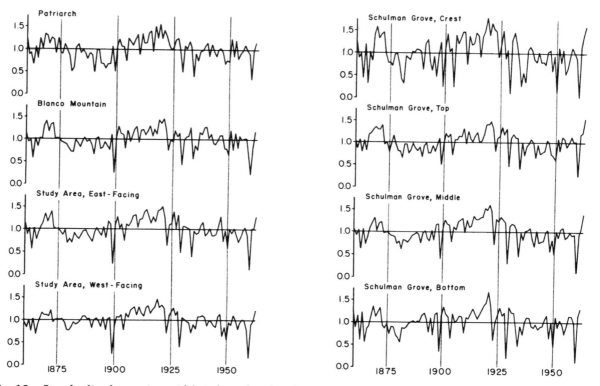

Fig. 12. Standardized tree-ring width indices for the chronologies of eight additional bristlecone pine stands in the White Mountains.

Fig. 13. The Schulman Grove Crest (site 15) which supports gnarled and old bristlecone pines with extremely variable ring-width chronologies. Note the rocky soil and wide spacing of trees which contrasts with the dense forest cover on the north-facing slope shown on the extreme right.

stands, trees are especially dense. The tree-ring characteristics of west-facing and east-facing slope samples are similar in other respects. Characteristics are intermediate between the old trees in the valley and those on the south-facing, mid-slope sites.

Variation Due to Slope Aspect on the Schulman Grove Sites

Tree-ring characteristics along a similar topographic gradient but with southeast rather than a southwest exposure are shown in the next four rows of Table 7 (sites 15, 13, 14, & 16), and the chronologies are plotted in Figure 12. The crest site (15) is on a north-south ridge (Fig. 13). The stand on the crest and the stand on the lower slope (site 16) both are exposed to the predominantly west winds sweeping through the valley. The upper and mid-slope

stands (sites 13 & 14) are below the ridge on the protected leeward and southeast-facing slope.

The statistical characteristics for the crest sample indicate the extreme aridity of that site. Rings are narrow, many are locally absent, and widths are highly variable from year to year. Correlations among cores, among trees, and with other groups are high; and the percent variance in the group chronology is high. The percent of variance attributed to cores is smaller than in any other site.

The characteristics for stands on the slope below the crest exhibit a gradient which is the reverse of that for the south-facing slope in the Study Area sites. The trees on the upper slope below the crest exhibit somewhat wider rings, fewer missing rings, less chronolgy variation, higher serial correlations, and relatively low correlation among cores and trees.

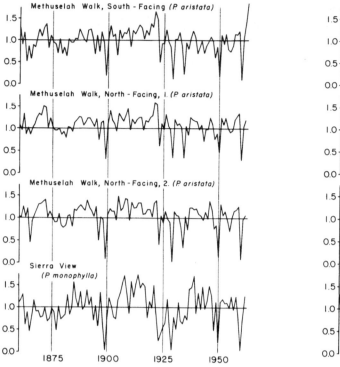
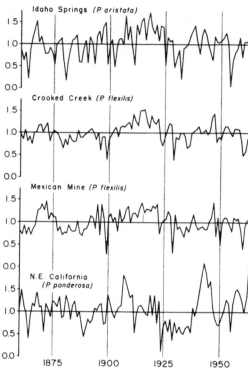

Fig. 14. Standardized tree-ring width indices for six stands representing pine species from the White Mountains, California, one bristlecone pine stand near Idaho Springs, Colorado, and a ponderosa pine stand in Northwestern California.

The rings from trees on the lowest site (16) exhibit more variability, they are more frequently missing, and there is somewhat more consistency among the tree chronologies within the group. It is inferred from these data and from observations of aerial photos made in spring (LaMarche, personal communications) that drifting snow accumulates in the lee of the crest and supplements soil moisture in the upper and to a less extent the middle-slope sites.

Variation Due to Elevation on South-Facing Slopes

Sites 1, 14, and 17 (lines 13-15, Table 7) represent an elevational gradient of south-facing slopes from the Patriarch area (high) to the Methuselah Walk area (low), an elevation difference of approximately 1,100 feet (335 m) (Figs. 12 & 14). The mid-slope in the Study Area (Site 7, line 4) and lower slope of the Schulman Grove (site 16, line 12) may be added as intermediate and moderately low elevation sites. The mean ring width does not vary markedly with elevation, but rings are more frequently absent, the stand is less dense, and trees more contorted in the

lower elevation sites (Fig. 15). The chronology at low elevations is more variable (Fig. 14), there is less serial correlation, and the correlation among trees is high. There is also a high percent variance in the group chronology, and differences among trees and cores are low. Since stands 17 and 1 are not centrally located, their correlation with other group chronologies is not exceptionally high.

Variation Due to Substrate

Sites 3 and 5 (line 16 and 17, Table 7) represent a south-facing stand on granite (Sage Hen Flat, Fig. 11) and a west-facing stand on sandstone (Silver Canyon, Fig. 11). Both sites produce relatively large, widely spaced trees which exhibit substantially large rings. Rings are more commonly absent in trees on sandstone than on granite, but relative variability of ring-width patterns, serial correlations, and correlation within and among trees on the granite site is high. The percentage of variance in the group chronology for granite is larger, but the errors of individual indices are the same. However, the group chronology on sandstone is more highly correlated with trees on dolomite than is the chronology on the granite site.

Fig. 15. The Methuselah south-facing slope as viewed from the north-facing slope. Note the variable spacing and stature of trees which change markedly with small changes in exposure and slope.

Variation Associated With "Sensitivity" of Chronology in Old Trees

Site 20 (lines 18 and 19, Table 7) represents two samples of old-age and low-elevation trees obtained by C. W. Ferguson from the lower portions of the northward draining valley known as Methuselah Walk (Fig. 15). His paired cores were not related to crown density, and in some cases were from strip-bark rather than from full-bark trees. For purposes of analysis, he divided his samples into two groups of nine trees. One group represented the most variable (sensitive) chronologies and the other the least variable (complacent) chronologies (see Methuselah Walk, north-facing 1 and north-facing 2, Fig. 14). The two groups actually represent a gradient rather than two distinct types, as the trees were intermixed on varying microenvironments in the same site (Fig. 15). Ferguson's samples

represent the more optimum site conditions in the area, where he has utilized longer ring-width records exhibiting similar variability to establish a year-to-year control for a 7,104-year chronology (Ferguson 1969). He was able to obtain longer replicated cores, so he extended the basic length of his series to include 260 more years. Therefore, his replicated chronologies start in A.D. 1600. This sample provides a transition between the relatively young trees used for my statistical analysis and the much older trees used for building long chronologies. The data presented here are based upon index values for the period 1860 to 1962.

The differences between these samples for mean ring widths, percentage of missing rings, standard deviation, and mean sensitivity reflect Ferguson's division into sensitive and complacent groups. The serial correlation in both groups was low, but correlations within and among trees for the complacent sample are lower than for any other group. The sensitive group exhibits higher correlations and a higher percentage of chronology variance than does the complacent set, but these parameters are generally lower than the median values for other sites.

Variation Among Species

Sites 2, 18, and 19 (the last three lines in Table 7) list the ring characteristics for two samples of limber pine and a sample of young but extremely high-elevation pinyon pine (St. André et al. 1965). These data show that the chronologies for limber pine (Fig. 14) are similar to those of the mesic-site bristlecone pines. On the other hand, the chronologies from pinyon pine (Fig. 14) are highly variable and many rings are absent. Correlations for cores within trees are high, but the correlations among trees and the percentage of variance in the group chronology are low. Additional tree-ring samples from pinyon pine at lower elevations exhibited greater variability in ring widths, and more rings were absent, to the extent that tree-ring dating was either extremely difficult or impossible.

The data in Table 7 represent a wide variety of dateable ring-width chronologies in the White Mountain area. Dense stands of bristlecone pine on moist north-facing or west-facing slopes exhibit the least variability in ring chronology, and there are wide differences from radius to radius and from tree to tree.

On south-facing slopes, ridge crests, and at low elevations where sites are arid, ring-widths are more variable and local absence of rings is more likely to occur. The amount of variance in the common chronology is high. The most extreme environment is on windswept rocky crests which have little soil and which are blown free of winter snow. The crest-site chronologies exhibit high correlation with the other bristlecone pine and can be considered the most representative and most climatically "sensitive" site for bristlecone pine.

Relation to Chronologies in Other Areas

Comparative chronologies for a bristlecone pine from Colorado (Fritts 1965) and ponderosa pine (*Pinus ponderosa* Laws.) at lower elevations in northeastern California are shown in Figure 14. Although details of these chronologies differ from those for the White Mountain groups, the general chracteristics, especially for the Colorado bristlecone pine, are similar (Fritts 1965). The chronology for ponderosa pine is somewhat more variable and exhibits higher serial correlation.

Normality

A chi-square test for goodness of fit, as described by Arkin and Colton (1957), was used to test for the normality of the group chronologies. All chronologies were slightly skewed with a higher than normal number of observations in classes above but near the mean, and in the lowest class representing the narrow rings. The Methuselah Walk south-facing and Methuselah Walk "complacent" chronologies exhibit significant skewness ($0.95 < P < 0.99$). However, the Methuselah Walk "sensitive" and the Study Area north-facing top samples exhibited more skewness ($P > 0.99$).

CHRONOLOGY RELIABILITY AND SAMPLE SIZE

The error of individual indices (Table 7) is a reciprocal function of sample size. It is calculated from the analysis of variance data (Table 7) in the following way:

$$E_y = \sqrt{\frac{EMS_y \cdot PMS_t}{PMS_y \cdot N_t} + \frac{EMS_y \cdot PMS_c}{PMS_y \cdot N_t \cdot N_{ct}}}$$

where E_y is the error of individual indices; EMS_y is chronology variance component; PMS_y, PMS_t, and PMS_c are percent of variance components due to group chronology, tree chronologies, and core chronologies; and N_t and N_{ct} are number of trees and number of cores per tree. The error values expressed in Table 7 are for a standard sample of 10 trees, 2 cores per tree.

It may be noted from these data that the mean errors for the bristlecone pine chronologies shown in Figures 11, 12 and 14 range between ± 0.045 and ± 0.084. The greatest error is encountered in the variable chronologies on the most arid sites. It should be emphasized that this is error of estimate for the chronology index, not error in dating, that is, the assignment of calendar year to each annual ring.

The mean square components (EMS) in analyses of variance are independent of sample size (that is, are

TABLE 8

The Calculated Number of Trees Which Must Be Sampled From Specified Sites to Assure That
the Group Ring-Width Chronology Contains an Error Less Than 0.08 or Less Than 1/4
the Standard Deviation of the Group Chronology*

Sample		Number of Trees								Ratio
		Error of 0.08				Error of ¼ the St. Dev.				
Site No.	Name	Radii sampled/Tree				Radii sampled/Tree				Tree Variance/ Core Variance
		1	2	3	4	1	2	3	4	
8	Study area, young	9	6	5	4	19	11	9	8	0.32
8	Study area, old	9	6	5	5	15	10	9	8	0.61
6	Study area, south-facing, top	6	4	4	4	8	6	5	5	0.79
7	Study area, south-facing, middle	6	4	3	3	10	7	6	5	0.56
9	Study area, north-facing, middle	7	5	4	3	19	11	9	8	0.30
10	Study area, north-facing, top	9	7	6	5	12	9	8	7	0.80
11	Study area, west-facing	5	3	3	3	11	8	7	6	0.72
12	Study area, east-facing	6	4	4	3	11	8	6	6	0.54
15	Schulman Grove, crest	10	8	7	6	8	6	5	5	1.06
13	Schulman Grove, top	7	5	4	4	13	9	8	7	0.70
14	Schulman Grove, middle	10	6	5	4	16	10	7	6	0.20
16	Schulman Grove, bottom	8	6	5	4	12	8	7	6	0.55
1	Patriarch	7	5	4	4	13	9	7	6	0.48
4	Blanco	7	5	4	4	13	9	8	7	0.57
17	Methuselah Walk, south-facing	8	6	5	5	10	7	6	6	0.81
3	Granite	8	6	5	4	10	7	6	5	0.63
5	Sandstone	8	6	5	5	16	12	10	9	0.79
20	Methuselah Walk, complacent	13	8	6	6	23	14	11	10	0.29
20	Methuselah Walk, sensitive	12	7	5	5	15	9	7	6	0.19
2	Limber Pine	6	4	4	3	14	10	8	8	0.66
18	Limber Pine	9	6	5	4	15	10	8	7	0.47
19	Pinyon Pine	23	16	14	13	17	12	10	10	0.69

*Calculations are based on the tree and core variances (Table 7) and are presented for cases where 1, 2, 3, and 4 cores are sampled from each tree. The ratio of tree variance to core variance is shown on the right.

estimates of the population mean). Therefore, selected levels of error can be specified, and by using the estimated mean squares for each sample in the above equation, the numbers of trees and cores needed from each stand to reduce the error below the specified level may be obtained. Such data enable the dendrochronologist to evaluate his sampling design.

Table 8 presents the results from calculations using data from Table 7. The specified error is set at 0.08 and at a value proportional to the standard deviation of each chronology. Calculations of number of trees required to achieve the error levels are shown for samples of 1, 2, 3, and 4 radii per tree. The ratio of variance due to tree versus variance due to cores shown in Table 8 is an index expressing relative importance of these two components. In general, the variance among trees is greater so the ratios are less than one. The table shows that an error of 0.08 using a sample of one core per tree could be obtained by sampling as few as 5 to 13 trees depending upon the site. If two cores were obtained from each tree, it would be necessary to sample only 3 to 8 trees, depending on site. However, the second sampling scheme would require coring a total of 6 to 16 radii (2 cores per tree) or 1 to 3 additional cores per site. Calculations based upon other specified errors show similar results. These data would indicate that an extensive sample of one core from many trees may provide more information for the same expenditure of time than an intensive sample of several radii from only a few trees within a site.

Calculations using specified errors which are one-quarter of the chronology standard deviation (Table 8) show that relatively reliable chronologies may be obtained from a small number of single-core samples from arid-site trees (sites 6, 7, 15, 17). Additional cores from the same trees add little information because the core variance due to differences within trees is small (the tree-core variance is large). However, for stands on the north-facing and protected sites (20, 14, 9) and for the young tree sample (8), the core variance is three to five times as large as the tree variance (ratios of 0.30 to 0.19). In these cases, an additional core (two cores per tree) may contribute sufficient new information to justify a replicated sampling design. Further replication (3 and 4 radii) would appear to be an inefficient way to improve chronology reliability, except where only a few trees are available for a given site. Chronologies representing early periods of time may be derived from a limited number of old trees, and in this case the chronology error may be reduced somewhat by measuring additional radii from the available trees.

DENDROCLIMATIC RELATIONSHIPS

The initial regression analyses were run only on samples from sites 7, 10, 11, 12, 13, and 14. Six separate series of analyses were obtained for each site, using the following independent variables:

1. Total precipitation for July through November, for December through March, and for April through June prior to growth, and the three prior ring-width indices.
2. Total precipitation as above but omitting the three prior ring-width indices.
3. Average temperature for the same periods as in series 1 including the three prior ring-width indices.
4. Average temperature as in series 3 but omitting the three prior ring-width indices.
5. Total precipitation as in series 1, average temperature as in series 3, the product between precipitation and temperature for each seasonal period, and the three prior ring-width indices.
6. Same as in series 5 but omitting the three prior ring-width indices.

The three prior ring-width indices allow for serial correlation to be taken into account, and the cross-products allow for the assessment of interaction (or non-additive relationships) among the independent variables.

When only precipitation and the three preceding tree-ring indices were used as independent variables (series 1), multiple correlations ranged from 0.44 to 0.74. When only temperatures and the preceding tree-ring indices were used (series 3), multiple correlations ranged from 0.55 to 0.73. When both precipitation and temperature were included along with their products (series 5), multiple correlation ranged from 0.67 to 0.79. Only the ring-width index for the third year prior to growth was entered, and its coefficient was negative.

The same six samples were screened further in a series of analyses using the following independent variables:

7. Total evaportranspiration deficit for April through May, for June, for July, and for the first half of August, and three prior ring-width indices.

8. Total evapotranspiration deficit as in series 7, omitting the prior ring-width indices.

9. The same variable used in series 7 plus the square of total evapotranspiration for April through May, for June, for July, and for the first half of August.

10. The same variables as in series 9, omitting the prior ring-width indices.

In analyses 7 and 8 only June evapotranspiration deficit was significant, giving correlations from 0.61 to 0.68. In analyses 9 and 10 the squared term for

TABLE 9

Climatic Conditions Correlated With Narrow Rings of Bristlecone Pines in the White Mountains*

Sample		Season Interval			
No.	Name	Previous July - Nov.	Dec. - March	April - June	Multiple Correlation
8	Study Area, young	cool†		dry warm†	0.958
8	Study Area, old	dry	dry cool‡	dry warm†	0.959
6	Study Area, south-facing, top		moist	dry warm†	0.968
7	Study Area, south-facing, middle	dry	cool‡	dry warm†	0.969
9	Study Area, north-facing, middle			dry warm†	0.854
10	Study Area, north-facing, top	dry warm‡		dry warm†	0.964
11	Study Area, west-facing		cool	dry warm†	0.925
12	Study Area, east-facing	dry warm		dry warm†	0.934
15	Schulman Grove, crest	dry‡	cool	dry warm‡	0.944
13	Schulman Grove, top	dry	cool	dry warm†	0.865
14	Schulman Grove, middle		cool‡	dry warm†	0.940
16	Schulman Grove, bottom	dry†	cool moist†	dry warm†	0.975
1	Patriarch	dry warm‡	cool†	dry warm†	0.958
4	Blanco Mountain	dry warm†		dry warm†	0.916
17	Methuselah Walk, south-facing	dry		dry warm†	0.952
3	Sage Hen Flat, granite			dry warm†	0.730
5	Silver Canyon, sandstone			dry warm†	0.832
20	Methuselah Walk, north-facing†		cool	dry warm†	0.915
20	Methuselah Walk, north-facing‡	dry	cool‡	dry warm†	0.971
2	Crooked Creek, limber pine			dry warm‡	0.621
18	Mexican Mine, limber pine	dry warm†	dry†	dry warm†	0.983
19	Sierra View, pinyon pine	dry warm‡	dry warm†	dry warm†	0.975

*Regression analysis terminated when F < 2.5.
†At least one coefficient significant at 0.99 confidence level.
‡At least one coefficient significant at 0.95 confidence level.

June was consistently entered for five of the six samples, and prior indices for 3 samples. Multiple correlations ranged from 0.65 to 0.86. Without the previous indices (Series 10) multiple correlations ranged from 0.65 to 0.74.

Serial correlations in the White Mountain ring series are generally small and are positive (Table 7), whereas previous growth was selected in the above analysis to be inversely related, accounting for a small variance. Because of this contradiction, it was concluded that the apparent inverse regression with prior ring indices was spurious, and the ring indices for preceding years were no longer used.

The final model that was tested employed temperature and precipitation data and their products for the July through November and the December through March periods. However, total evapotranspiration deficits, their squares, and their cubes were used for the months of April, May, and June that immediately preceded growth. The regression equations that were obtained were solved for selected values of precipitation, temperature, and evapotranspiration deficit following Fritts et al. 1965a. Table 9 summarizes these results in word form showing the conditions associated with narrow rings and the corresponding multiple correlations. Even though there were only 15 years of data, the degrees of freedom for the equations used in the table were never less than eight. All multiple correlations were significant from zero.

These results demonstrate that for all 22 samples the climate for April through June, as measured by calculated evapotranspiration deficits, is most highly correlated with ring-width growth. In 13 cases dry or dry-warm climate during the previous July through November is associated with low growth. Temperature for July through November and ring widths were directly correlated only in the case of the young trees from site 8. For 10 samples low temperatures during December through March were significantly associated with low growth. The most highly significant relationship between winter temperature and growth was obtained in the analysis of the high elevation trees (Sample 1, Table 9). Only in the low elevation limber pine and pinyon pine was low moisture during the winter related to low growth.

Differences in the multiple correlation coefficients in Table 9 appear related to certain characteristics of the sampled sites. The bristlecone pine on Sage Hen Flat and limber pine at Crooked Creek, both on granite soils, were the only samples exhibiting multiple correlations below 0.8. The north-facing stand on the mid-slope of the Study Area, the stand on the

upper snow accumulation slope in Schulman Grove, and the stand on the Silver Canyon sandstone were the only groups with multiple correlations between 0.8 and 0.9. The highest multiple correlations were obtained in the analysis of the arid-site stands growing at low elevations, on windswept ridges, or on south-facing slopes.

Three of the regression lines for ring indices expressed as a function of evapotranspiration deficit in June are plotted in Figure 16. The points represent the original observations. In each case a third-degree polynomial was required. Although some of the departure of the observations from the regressions was shown to be related to climate for the previous July through March, it is evident that growth is a highly non-linear function of June evapotranspiration

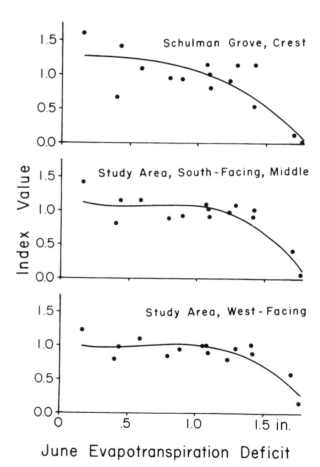

Fig. 16. The calculated total evapotranspiration deficit for June plotted against three tree-ring chronologies from the White Mountains. The curve is a third-degree polynomial fitted to these data by the stepwise multiple regression analyses. Departures of points from the regression include the effect of factors during other seasons as well as the statistical error.

deficit. Thus during those years when evapotranspiration deficits in spring are large, growth on all sites is severely reduced. At other times a variety of less influential factors may become limiting to growth. This would produce the observed skewed bimodal population with one group of very narrow rings which are easily distinguished in trees on all sites, and a second but larger group of average-size rings which are less diagnostic and which may be limited by different environmental conditions on various sites.

CLIMATE AND ELONGATION OF NEEDLES

Measurements from the branches and needles of four bristlecone pines sampled from the Schulman Grove crest site were evaluated (Table 10) using analysis of variance (Snedecor 1956). Length of needles was the only characteristic which exhibited significant variation from year to year. The annual mean needle lengths for 15 years amounted to 19% of the total variance. The individual trees exhibited 44% of the variance, while interactions of years with trees and the two levels of branches with trees accounted for 10% and 1%.

The mean needle lengths and climate for the 15 years were then analyzed by multiple regression following the procedures used for ring-width chronologies. Needle lengths are directly correlated with April through June precipitation and inversely correlated with December through March precipitation. The multiple correlation was 0.71. Needle lengths were also found to correlate with the square and cube of evapotranspiration deficits for April through June. These variables for April through June were then entered into a multiple regression with temperature and precipitation for December through March. The variance accounted for was 81% (R = 0.90). As was the case in analysis of ring-widths, the climate prior to but not during the growing season is statistically related to needle-length growth.

TABLE 10

Analysis of Variance in Needle Lengths of Bristlecone Pine Representing Growing Seasons of 1949 Through 1963

Source	df	SS	MS	EMS	%EMS
Years	14	574.25	41.02*	3.91	19
Levels	1	1.01	1.01	0	0
Trees	3	841.42	280.47*	9.02	44
Years x levels	14	73.37	5.24	0	0
Years x trees	42	408.83	9.73	2.14	10
Levels x trees	3	26.84	8.95	.23	1
Years x levels x trees	42	228.78	5.45	5.45	26
Total	119	2154.50	..	..	..

*Highly significant at 0.99 confidence level.

DISCUSSION

FACTORS AFFECTING RING-WIDTH GROWTH

This study has demonstrated that there are many interacting factors which control the growth of a tree ring in the White Mountains. However, only those factors which directly or indirectly limit the rates of growth processes, such as cell division, cell enlargement, and cell differentiation, will influence the structure and character of the annual ring (Blackman 1905, Mitscherlich 1909). Therefore, the following discussion will treat first the possible limiting environmental factors that immediately affect conditions of the growing tissues within the plant. Some of these internal conditions are directly or indirectly coupled to factors of the external environment. Others result from changes in related processes which in turn are affected by environment.

However, most of the important environmental factors do not have an immediate effect. These may be represented as conditions of the past environments or changes in processes, structures, or substances which may influence growth. These relationships are grouped for convenience under history of environment.

A third section will consider biotic interactions that occur within the tree and among the organisms of the forest community.

The last section will treat the modifying effects of site which influence the range and variability of environmental factors creating microenvironmental conditions which affect growth.

Immediate Environment

The growth processes governing the formation of a ring may be directly affected by moisture as it influences the cell water balance, temperature as it affects rates of biochemical processes, day length and temperature as they influence length of growing period, and wind and temperature and other disturbances as they may injure the actively growing cell.

A. **Cell water balance.** Active cell division and rapid cell enlargement are both dependent on an adequate cell water balance. A reduction in the moisture absorbed through the roots or an increase in the water transpired from the leaves may produce water deficits within the tree. The apparent correlation between precipitation and widths of rings from arid-site trees has often been attributed to the direct

action of the environment on the water balance which by inference becomes directly limiting to the growth of a tree (Glock 1955, Glock et al. 1964, Glock and Agerter 1966).

The dendrograph records and environmental measurements of our study indicate pronounced changes in the water balance of bristlecone pine during the three seasons of growth. In spite of these differences, the cambium appeared active throughout both wet and dry climatic extremes. The drought in 1964 appeared to reduce cell enlargement, but in 1962 there was no perceptible change associated with the soil moisture regime. During 1962, rainfall in late summer may have prolonged growth in certain young trees on warm sites, but it failed to alter the growth rates in many of the older trees. Regression analyses have repeatedly failed to establish any statistical relationship between tree-ring width and climate for the growing period.

The almost complete absence of intra-annual latewood bands or false rings in bristlecone pine is additional evidence that changes in midsummer climate do not markedly influence the structure of the current season's ring. These results are in contrast to the findings of Glock et al. (1960) and Glock and Agerter (1963) where branches from a wide variety of Texas-grown trees reveal multiple growth layers attributed to varying temperature and soil moisture regimes. However, their studies utilized young branches from low elevation trees which, in contrast to White Mountain species, may grow several times during a long frost-free season. The occurrence of multiplicity in young trees at low elevations led Libby (1963) to improperly infer that discrepancies between tree-ring and radiocarbon dates in high-elevation bristlecone pine may be attributed to frequent double rings. All studies that have been conducted in the White Mountains indicate that distinct double rings rarely occur.

B. **Temperature of the growing tissue.** Both the rates of respiration and assimilation of cell materials may be a function of temperature. At high altitudes plant temperatures during the growing season may be sufficiently low to directly limit growth. Figure 6 shows that the rate of xylem production was lower in the cool late June and early July of 1963 and 1964 than in the comparable 1962 period. During 1963 the initiation of cambial activity was further delayed by 10 to 12 days in the higher elevation sites.

However, the dendrometer, phenological, and anatomic measurements indicate that higher rates of growth occurred during subsequent warm periods. Although growth in 1963 started later, it also continued later, so that the length of the season was longer by two to three days.

Since mean monthly temperatures for all three of the growing seasons were below normal (Fig. 5) and no marked direct correlation of summer temperature with the ring-width chronology was obtained, it may be concluded that on the sites we studied, temperature during the growing season does not become sufficiently limiting to the total amount of cambial activity to significantly affect the width of the annual ring.

C. **Day length.** The initiation and cessation of cambial activity is sometimes controlled by seasonal changes in day length (Kramer and Kozlowski 1960). The beginning of the growing season in the White Mountains appears to depend upon favorable temperature (Fig. 6), and cambial activity ceases in approximately 45 days, independent of either day length or temperature. Day length does not appear to control the growing season and apparently has no influence on varying widths of annual rings.

D. **Damage to cambial tissue.** It is well known that the wounding of growing tissues may kill the cambium or may distort or accelerate growth, thus possibly resulting in an abnormal ring (Kramer and Kozlowski 1960). During the course of this study, sequential samples for anatomic plugs and dendrometric measurements revealed several distortions attributed to wounding at a previous time. Occasional rings indicated injury from fire, lightning, animals, or falling rocks. Glock et al. (1960) describe distinctive frost rings caused by severe freezing during the period of cambial growth. While no frost rings were observed in our study, they have been reported for higher elevation bristlecone pine (Ferguson and LeMarche personal communications).

Such injuries to the cambium are uncommon in arid-site White Mountain trees. Replicated samples for ring-width chronologies minimizes further the effect of these occasional damaging factors, though they undoubtedly contribute to a portion of the statistical error.

History of Environment

The results of this study indicate that the largest effect of environment on growth is not immediate.

Regression analyses show a correlation between ring widths and moisture conditions of the previous summer and autumn, temperature of the winter, and evapotranspiration deficits during April, May, and June, which all precede the beginning of growth. These environmental conditions may directly affect certain plant processes such as photosynthesis and respiration which control the accumulation of food and other materials necessary for growth. They also may modify other environmental conditions such as soil moisture which directly or indirectly influence growth.

A. **Temperature.** The processes of respiration and photosynthesis may at times be largely dependent on variations in the temperature regime (Kramer and Kozlowski 1960). Schulze et al. (1967) and Mooney et al. (1966) have presented data on rates of photosynthesis and respiration as a function of temperature for White Mountain bristlecone pine during June, August, November, January, and April. They found that relative rates of net photosynthesis at comparable temperatures was highest in June and lowest in January and April.

In addition to a greater photosynthetic efficiency during the month of June, the moderate day temperatures and cool night temperatures (Fig. 6) would result in rapid accumulation of photosynthate (Mooney et al. 1966) just prior to growth. Regression analyses of ring-width variation and climate suggest that after initiation of cambial activity which occurs late in June, a large portion of the currently produced photosynthate may be directly utilized in meristems within the tree crown; therefore, ring growth at the stem base probably depends on reserves of stored food produced during earlier periods (Kramer and Kozlowski 1960, Kozlowski and Keller 1966 p. 352, Fritts et al. 1965c, Fritts 1966). Also, the regression analyses of needle lengths indicate that growth in the crown may be partially dependent on food accumulated during the previous winter.

However, during late summer and autumn, after meristematic activity has ceased and day and night temperatures are reduced, there may be a net gain in stored food even though the efficiency of the photosynthesis may be somewhat less than in June. This food can be utilized in the next year's growth.

Schulze et al. (1967) have reported higher rates of respiration than photosynthesis during the winter months, and they attribute these rates to low winter temperature and the destruction of chlorophyll. Fritts (1966) has reported similar winter-time reduction of photosynthesis for ponderosa pine, which is attributed to freezing in the stem. The ice blocks water transport and produces water deficits in leaves which are subjected to a relatively high heat load during clear days.

The severity of winter temperatures may directly correlate with the inactivation of the photosynthetic mechanism in these high altitude trees. Cold winters may result in a prolonged inactive period and in a depletion of food reserves (Schulze et al. 1967). A warm winter may result in a net gain or in a less rapid depletion of reserves.

While temperatures during the winter may directly affect food reserves and indirectly limit growth, temperatures at other seasons interact with the moisture regime to produce differences in net photosynthesis, food reserves, and ring-width growth.

B. **Moisture.** While the water balance within the tree may rarely become directly limiting to cambial activity, there is evidence that it has a pronounced effect upon photosynthesis and therefore indirectly affects the growth of a tree.

Both Gates (1965) and Fritts (1966) have presented evidence that water deficits may drastically reduce the cooling power of transpiration and induce elevated temperatures in leaves. Measurements made in this study show that daytime air temperatures frequently exceed those that are optimum for photosynthesis (Mooney et al. 1966). Water stress as measured by June evapotranspiration deficits may induce even higher needle temperatures, so that net photosynthesis during mid-day may be significantly reduced.

Moisture deficits may directly limit net photosynthesis in bristlecone pine. Wright and Mooney (1965) produced severe reductions of net photosynthesis in potted plants by reducing soil moisture to levels near wilting percentages. At all other soil moisture levels the rate of photosynthesis remained high.

The regression analyses are best interpreted as the interaction of the annual temperature and moisture regimes on the accumulation of stored foods which ultimately control the rate of cell production and the relative width of the annual ring. Drought during the most active photosynthetic period, May through June, apparently becomes limiting to photosynthesis. Since the most food is produced at this time, drought in May and June is most highly correlated with narrow rings. Drought in late summer and autumn may have a lesser effect upon the food reserves and therefore is less highly correlated with narrow rings.

C. **Wind.** Wind is an important factor especially during winter as it removes snow from west-facing or exposed open slopes and deposits it in valleys, behind barriers, or in the dense forest stands. This results in a mosaic pattern of sharply contrasting soil moisture regimes (Wright and Mooney 1965). Wind-carried snow may blast and destroy tree foliage or bark on exposed sites (Fig. 13). The destructive effects of wind are most dramatically shown by the old and gnarled trees growing on the exposed ridges and crests. Wind will also sweep away warm and moist air surrounding the tree crown and modify the water and energy balances (Gates 1965).

Biotic Factors

The effects of environmental factors are highly dependent upon biotic interactions within the tree and between the individuals of the forest stand.

A. **Interactions among growth processes within the tree.** Growing shoots and roots, as well as developing fruits, may compete with the cambium for food, may affect cell formation through the production of growth regulators, or may change the water balance of the tree. The phenological observations on bristlecone pine show close correlations and timing among all growth processes during all three years (Figs. 6 & 7). Bud swelling just precedes cambial growth. Early in the season both terminal and cambial growth may be slow. By midseason bud elongation occurs and is followed by needle emergence and subsequent growth. Pollination occurs at the end of the growing season as the needles reach their mature size, and the last cells of the annual ring are formed.

Female cones may compete for food and reduce the growth in the near-by cambium. The rings of twigs were observed to be narrowest near the attachments of the developing cones. However, no relationship could be found between cone production and the ring widths at the base of trees.

Needle growth, like cambial activity, varies from year to year; but these needles are retained and apparently remain active from 15 to 30 years. Hence the total needle mass and potential photosynthetic capacity for any one time represents the accumulated climatic effects on needle production over a large number of years. This accumulated needle mass does not vary significantly from year to year but may respond to only long-term climatic change. Thus an extreme drought for one year is reflected by a single narrow ring and not by reduced growth in subsequent years. First-order serial correlation in ring widths is therefore low. It is probable that this relatively stable needle mass and the associated low serial correlation is an important factor contributing to the ability of bristlecone pine to withstand the extreme adversity and variability of exposed and semiarid high-elevation sites.

B. **Physical condition and structure of the tree.** The needles of bristlecone pine may vary in color, shape, and structure, which may affect different processes in the tree. During winter, needles turn yellow and photosynthesis cannot occur (Schulze et al. 1967). During the summer, variations in color and moisture may affect the absorption and dissipation of energy and influence temperatures within the tree.

Young trees are less exposed to wind and are more subject to shading than the older and larger trees. Cambial growth may start earlier, end later, and occur at a faster rate in young trees (Tables 5 & 6). As a result, young trees have wide rings, few rings are absent, and the variance in the chronologies is less well correlated within, as well as among, trees.

On exposed dolomite slopes many bristlecone pines become stunted and deformed. Portions of the crown and cambium may die, new leaders may develop, and only a small strip of bark connecting the live top and roots may survive. This so-called "strip-bark" tree apparently reaches a quasi equilibrium with its environment.

Since photosynthetic efficiency decreases as the trees grow old (Wright and Mooney 1965), dieback of the nongreen tissues in the trunk and root helps maintain an equilibrium between the food-making and food-using tissues. This feature enables the tree to persist and grow for many years, reaching ages over 4,000 years. Rings are so narrow that any gradual reduction or increase in the total photosynthetic area of the top may be accompanied by the reduction or expansion of the cambial tissues rather than a change in the width of the rings. As a result, the rings from old semiarid-site bristlecone pine provide uniquely homogenous, millenia-long time series.

The chronologies from strip bark trees exhibit high correlations with climate and with the chronologies of other trees; yet they are relatively free of long-term changes associated with increasing age. A high resin content preserves the wood, so that many old wood remnants may be found as partially buried fragments on the dry dolomite sites. These fragments may be cross-dated by using the inner portions of living trees, and the tree-ring chronology can be extended into the past by incorporating the remnants from successively older trees (Ferguson 1968).

On moist sites, especially on sandstone and granite, growth is rapid, and heart rot may develop in 16- to 20-inch diameter trees (Wright and Mooney 1965). Death occurs at a relatively young age, and the trees soon topple and rot away.

C. **Genetic potential.** Little is known about genetic diversity in bristlecone pine. However, the longevity and low reproduction rates would not favor rapid genetic change. No evidence was found for marked hereditary differences among trees. Comparisons among the chronologies of bristlecone, limber, and pinyon pines in similar sites show fewer differences than are found among the chronologies of the same species on contrasting sites.

D. **Competition within the forest stand.** Trees in the most arid and exposed sites are widely spaced so that intraspecific competition is not high. There is little seedling establishment even during moist years. The uniform distribution among size classes (Table 4) suggests that after establishment death due to competition rarely occurs. Seedlings occasionally are found near larger trees, but a high light requirement (Wright and Mooney 1965) precludes their growth in shade. Terminal buds are often killed, lateral branches will grow, and trees with multiple stems may form.

Trees on moist sites form denser stands, they are taller, and crowns are irregular as competition within and among trees is high. The width of a ring may vary within the same tree so that the percentage of variance in the chronology which is attributed to core and tree differences is high.

Modifying Effects of Site

Site factors in the White Mountains are important, as they modify the environment which indirectly or directly influences the width of the ring. The replicated tree-ring samples (Fig. 1) were chosen so that they may be grouped to evaluate single site variables (Fig. 17) and their statistical parameters compared. The relative change in these parameters is used to assess the effects of topography, elevation, substrate, chronology selection, and species on the ring-width characteristics of the trees (Fig. 17). With the exception of serial correlation, high values of each parameter indicate highly limiting environments (Fritts et al. 1965b).

A. **Topography.** The plots in Figure 17 show that the greatest changes in ring characteristics of bristlecone pine are related to the topographic differences in the site. In the valley oriented in the north-northeast and south-southwest direction (site 6-10), topography modifies incident radiation, temperature, wind, and soil moisture which may become indirectly limiting to the growth of trees. On the dry south-facing exposures and the steep and exposed upper north-facing slope, many rings are absent, their indices are variable, the patterns in ring-widths correlate highly within and between trees, and the variance component in the chronology is high. On the moist cool lower north-facing slope and on the valley floor (sites 9 and 8), where denser stands of bristlecone pine occur, chronologies are less variable, fewer rings are absent, correlations within and between trees are low, and the component variance due to chronology is less. Environmental factors are not as limiting to trees in these moist sites so that growth does not vary as markedly from year to year. Competition between trees and within the crown cause the high variance attributed to core and tree components and produce some long-term trends as indicated by the first order serial correlation which is proportionately high.

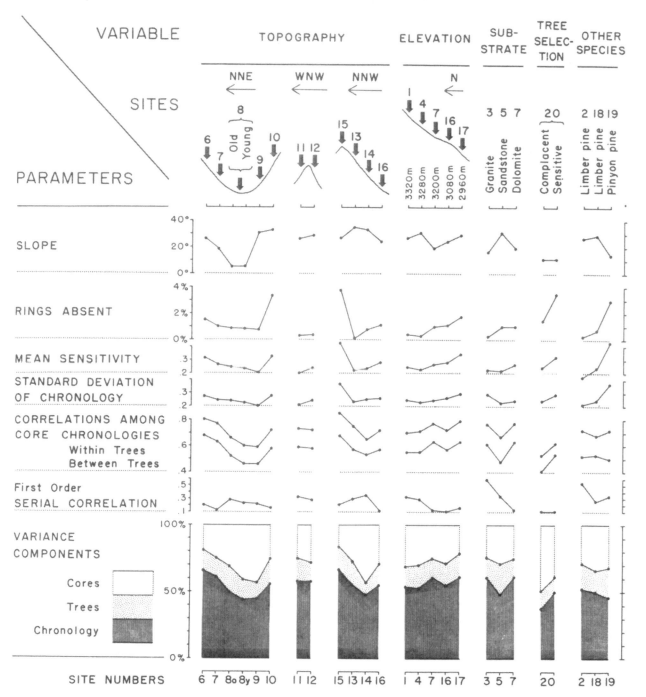

Fig. 17. Characteristic tree-ring and site parameters plotted as a function of topography, elevation, substrate, tree selection, and species. Data are from Table 7 and numbers for sites used for each variable are identified in Figure 1.

The two stands shown in the WNW sequence exhibit less marked chronology differences, though stand 11, which faces a windswept valley, shows some effects of drifting snow. Among these effects are wide rings (Table 7), somewhat less chronology variability, higher serial correlation, and more differences associated with the variance components of trees.

The tree-ring characteristics of the NNW series in Figure 17 indicate a highly limiting environment on the rocky crest (site 15, see Fig. 13). In contrast, characteristics of tree rings on the SSE-facing slope immediately below the crest (site 13) indicate the environment is much less limiting to growth. Few rings are absent; ring-widths are less variable; correlation within and between trees is low; serial correlation is high; and the percentage of variance due to chronology is less. It is inferred that drifting snow accumulates on the lee of the crest so that soil moisture is high. In contrast to the ring chronologies from other south-facing but drier windswept sites, the ring chronology from this snow accumulation area contains less information on variations in climate.

Large differences in tree-ring characteristics associated with variation in topography are evident. Slope, exposure, and orientation to prevailing wind are prime factors controlling moisture relations and the ring-width patterns in bristlecone pine.

B. **Elevation.** The parameters for ring characteristics of bristlecone pine along the elevational gradient (Fig. 17) indicate the environment is most limiting to ring-width growth at the low elevation sites. Air temperatures are higher, precipitation is less, snow melts more rapidly, and the heat load and evapotranspiration rates are high.

There is less change in the tree-ring parameters associated with elevation differences than with the topographic differences of site. But at low elevations the most marked change in vegetation and tree-ring characteristics associated with topography are observed. Subtle differences in relief and exposure produce marked changes in stand density and composition (Fig. 15) and also produce large variations in the chronologies within and among trees.

C. **Substrate.** As differences in parent material affect soil color and temperature, mineral content, and soil development, they may also modify the degree to which the environment limits processes within the tree. Wright and Mooney (1965) state, "The greater development of bristlecone pine on dolomitic soils appears related primarily to the better water relations of these soils, tolerance of bristlecone pine for low nutrient availability, and the lack of competitors."

The characteristics of tree-ring chronologies associated with substrate differences are shown in Figure 17. The slope was steeper and more westerly oriented for the sandstone area, so the sites also varied in the effect of prevailing winds. In all probability this different exposure to wind influenced the tree and core variance components and the corelations within and between trees. However, the rings of trees on the granite and to a lesser extent on the sandstone soils are generally wider than those on dolomite, fewer rings are missing, and serial correlation is high.

* * *

It would appear from these data that yearly variations in moisture for sandstone and granite are less limiting to ring growth than on dolomite sites. However, Wright and Mooney (1965) and Mooney et al. (1966) point out that at the seedling level, the environment is more unfavorable to the establishment of bristlecone pine on sandstone and granite so that dense forests rarely occur.

Where trees become established on granite and sandstone, they are relatively large and fast growing. Stand densities are low. Since little competition exists, roots may be established throughout a large soil mass, and in many years ample moisture may be available to the tree. A large root and shoot mass is established during years of favorable climate, but during unfavorable years, the water balance and food reserves may be drastically changed. Thus, differences in climate may produce large changes in the root-shoot ratio of trees on granite and sandstone, change their growth potential, and cause the serial correlation of ring-width series to be high.

* * *

D. **Other factors associated with site.** Dendrochronologists frequently use the ring-width variability and cross-datability of their samples (correlation between cores) as an indirect measure of the degree with which environmental factors have limited tree growth (Schulman 1956, Fritts et al. 1965b). Ferguson sampled his trees (stand 20) from the mosaic environments of Methuselah Walk (see Fig. 15). Instead of inferring site differences from features of microrelief, he examined the cores and grouped them into two numerically equal "complacent" and "sensitive" classes. It may be seen from Figure 17 that in the "sensitive" class more rings are absent, correlations among and between trees are greater, and the percent of chronology variance is high. However, in all of his trees there are marked differences from one radius to the next, so that the chronologies correlate less well within trees than for most other bristlecone pine sites.

The chronology characteristics of species other than bristlecone pine are shown on the right (Fig. 17). The limber pines were growing on granite and sandstone, while the pinyon pines were growing at lower elevations on a south-facing slope. Comparisons of their chronology characteristics with bristlecone pine on similar granite, sandstone, or low elevation sites show no marked differences between species.

The extreme aridity of the pinyon pine site caused many rings to be absent and the ring-width variability to be high. The youthfulness of the trees and the adversity of the extreme upper limit for this species help account for the low percent of chronology variance and for the low correlation between trees.

The regression analyses in Table 9 show that winter drought is more highly related to the chronology of pinyon pine than to the bristlecone and limber pine growing on colder and higher elevation sites. It may be inferred that winter temperatures are warmer for the pinyon pine at lower elevations (see St. André et al. 1965) so that the photosynthesis is more likely to remain active during winter and can be limited by winter-time drought (Fritts 1966).

RING ABSENCE AND PRECISION OF THE DATED SERIES

Dendrochronology is a precise tool in that every growth ring is identified (cross-dated) as to the year in which it was formed. This cross-dating is accomplished not by ring counts but by matching patterns of wide and narrow rings. The technique is especially necessary for White Mountain trees because in extremely dry years certain areas in the cambium may remain inactive, and only part of a ring will form. However, for a relatively few years (1899, 1920, 1934, 1950, and 1960) many partial rings were formed, and these rings are most likely to be found in trees on certain sites only (Table 7).

If a dendrochronologist relies on one stem radius from a small number of trees to establish cross-dating, it is possible that the ring for one extremely dry year could be missing in all observed radii. In such a case, a portion of his chronology could be wrongly dated by a factor of at least one year. This problem is essentially eliminated by either sampling trees from certain sites where rings are least likely to be absent, or by sampling a larger number of trees.

For example, the growth ring for 1960 was absent from more of the radii sampled than was any other growth ring since 1859. The most extreme case was the exposed north-facing slope (site 10) where rings were found in only 2 of 20 radii or in 10% of the cores that were obtained. The older trees of the nearby valley (site 8) exhibited the 1960 ring in 36% of the cases, while the radii selected for "sensitivity" from Methuselah Walk (site 20) contained the ring only 28% of the time. If cross-dating were based only on such trees, it would be necessary to sample a large number of radii to ensure that each year is represented by a growth increment in at least one sampled core.

However, the 1960 ring was observed in 90% of the radii from the more moist and cool upper elevation stand (site 1), in 64% of the samples from young trees in site 8, and in 67% of the cases from the Methuselah Walk "complacent" trees. Accurate dating of annual ring series from such trees would require only a small sample size.

These data verify the practice of assuring precison of dating by first building a chronology upon relatively complacent, young, or fast-growing trees. After a sufficient sample size has been obtained to assure correct dating, the dendrochronologist uses this relatively "complacent" chronology to date increasingly sensitive ring-width series representing progressively drier sites. Once these more "sensitive" trees with many partial rings are dated, their ring-width chronology may be used to reconstruct and analyze the climatic fluctuations of the past.

RING-WIDTH PATTERNS AS INDICATORS
OF CLIMATIC TYPE

Glock and Agerter (1966) classify ring-width patterns into three types. (1) The *California Pattern* which is ". . . characterized by marked uniformity in thickness, by sheaths of xylem entire over the plant body, and by unity within the annual increment. . . ." "Sequences of the California type occur not only in regions of winter rainfall but also in high latitudes and high altitudes and in regions of uniformly abundant rainfall throughout the year." (2) The *West-Texas Pattern* which is ". . . characterized by marked variability in thickness, by sheaths of xylem partial to a varying degree over the plant body, and by multiplicity within the annual increment." It is associated with regions ". . . under a continental type of summer rainfall." (3) The *Northern Arizona Pattern* which ". . . refers to a sequence of growth layers characterized by groups of 'California' type alternating with groups of 'West-Texas' type." They infer that this pattern results from an alternation of the summer-dominant and winter-dominant precipitation regimes.

The White Mountain stands are in California and at high elevations, so they should clearly fall in the California Pattern with a winter precipitation regime. As Glock and Agerter would predict, multiplicity in ring pattern is not common in White Mountain trees.

However, the rings are anything but uniform in thickness, especially on the ridges and dry sites. While Glock and Agerter may be correct in associating ring multiplicity with a warm climate and summer precipitation, it appears that they have inappropriately included ring-width variability as an attribute of a warm climate and summer-dominant precipitation regime. It can be clearly shown that under both winter-dominant and summer-dominant precipitation regimes, ring-width variability is a function of site aridity (Fritts et al. 1965b).

Therefore, alternating periods of high and low ring-width variability, which are noted by Glock in northern Arizona and which are also found in White Mountain trees, do not necessarily represent shifts from summer-dominant to winter-dominant precipitation regimes. However, they do indicate that past climate has been more or less arid at different times and has produced more or less variability in the thickness of annual rings.

SUMMARY

The climate of the White Mountains within the elevational range of bristlecone pine can be characterized as cool and arid. Precipitation may vary markedly from year-to-year.

Bristlecone pine is virtually restricted to dolomite substrates. The soil is a minimal calcisol and root penetration is obstructed by a massive accumulation zone at an 18 inch (46 cm) depth. Rock fragments occupy a large part of the soil mass so that the soil moisture storage capacity is low.

The greatest stand densities of bristlecone pine at lower altitudes are obtained on north-facing slopes and on sites where the normal precipitation is supplemented by the accumulation of drifting snow.

Growth in bristlecone pine starts with swelling of the buds in late June after freezing night-temperatures cease to occur. Growth stops about 45 days later, shortly after pollination is complete, even though soil moisture and air temperatures may remain high. The timing of the phenological stages was observed to vary from site to site and year to year, differing by as much as 13 days. During years of sufficient moisture the growing season may be longest in young trees or in trees on south-facing slopes. In years of drought all trees produce narrow rings, or in extreme cases cambial activity may fail to occur. The environmental conditions during the growing season have little effect on cell enlargement and differentiation so that distinct multiple or false rings are extremely rare.

Differences in ring width are more dependent on variations in the rate of growth than on variations in the season length. High evapotranspiration deficits during April, May, and June prior to growth have a marked effect on the width of a ring. Drought reduces net photosynthesis, which limits the accumulation of reserve food. With less stored food, growth is slow and a narrow ring is formed. Dry conditions due to high temperatures or low precipitation of late summer or autumn may also reduce the amount of accumulated food and thus affect the width of the following season's ring.

In lower-altitude pinyon pine, dry conditions during the winter may also result in low growth. However, bristlecone and limber pine growing at higher altitudes where winter temperatures are low are adversely affected by colder than average winters. Chlorophyll apparently becomes inactive under low temperatures, and photosynthesis cannot occur, so that the food reserves by the end of winter may be low. Therefore, the ring-width chronologies of bristlecone and limber pine record the collective effect of late summer and autumn moisture, winter temperature, and especially May and June moisture upon the subsequent season's growth.

The relationship between evapotranspiration deficits in June and ring width is curvilinear. As a result, the ring-width indices are not normally distributed. Narrow rings are abnormally abundant, while markedly wide rings are extremely rare.

In trees on arid dolomite sites, green needles are retained by branches up to 30 years; so a reduction in needle tissue resulting from a one-year drought does not markedly affect the potential photosynthetic capacity of the tree. This residual needle mass minimizes the long-term effects of drought, and, coupled with the capacity of the cambium to die-back when conditions are unfavorable, it may enable bristlecone pine to withstand environmental adversity of the harsh high-altitude sites. This results in ring-width chronologies which are highly variable, but the first order serial correlation is low.

On sandstone and granite, stand densities are lower, growth is more rapid, and a larger shoot mass is formed. Severe drought may markedly alter the root-shoot ratio, reduce ring-widths for several years, or even kill the tree. As a result the first order serial correlation in the tree-ring chronology from these sites is higher, and old trees are rare.

The various tree-ring chronologies in all stands of the White Mountains are highly correlated with each other, especially in years of minimum growth. The statistical characteristics of tree-ring chronologies are a function of environmental modifications caused primarily by topographic variation, secondly by altitude difference, and thirdly by substrate conditions of each site. Ring-width chronologies from the trees on exposed, low-elevation, and rocky dolomitic sites are most variable. They correlate best with chronologies in other trees in other areas, and they contain the most information about variations in climate. However, these chronologies are not easily dated because many rings may be small, or in extremely dry years, the cambium may fail to grow and no ring is formed.

Tree-ring chronologies from higher elevations, north-facing slopes, and from areas of snow accumulation exhibit the least chronology variability and contain the least information on past climates. However, a distinct ring is formed every year, and accurately dated chronologies based on a few trees can be readily obtained. The long-lived "complacent" trees that grow on the north-facing slope of Methuselah Walk are especially suited for chronology development because the sequences of extremely dry years are faithfully recorded as narrow rings, but locally absent or partial rings rarely occur.

Variance analyses of replicated ring-width series show that there are generally more differences in the chronology between trees than within trees. Computations show that for a given number of sampled radii, the standard error of a site chronology is lowest where single cores are sampled from a variety of trees. However, when few trees are available as during early periods of time, replicated measurements along several radii within the same trees can to some extent reduce the chronology standard error, as well as help to assure that partial rings have not been missed.

Changes in both the relative variability in ring-width indices and the mean ring-width index may be used to evaluate changes in limiting conditions associated with certain changes in climate. While these high-altitude tree-ring chronologies may differ from those at lower altitudes in that some variance is directly correlated with temperature rather than precipitation during the winter period, the long chronologies from the lower forest border of bristlecone pine are in large part moisture dependent. If the chronologies are developed from climatically "sensitive" arid-site trees and appropriately dated, they can provide an accurate, extremely long and reliable record of moisture and temperature as it has varied in the past.

LITERATURE CITED

ARKIN, H., AND R. COLTON
  1957 Statistical methods. Barnes and Noble, New York. 226 p.

BILLINGS, W. D., AND J. H. THOMPSON.
  1957 Composition of a stand of old bristlecone pines in the White Mountains of California. Ecol. 38: 158-60.

BLACKMAN, F. F.
  1905 Optima and limiting factors. Ann. Bot. 19: 281-95.

CURRY, D. R.
  1965 An ancient bristlecone pine stand in eastern Nevada. Ecol. 46: 564-66.

CURTIS, J. T.
  1956 Plant ecology workbook. Burgess, Minneapolis, Minnesota.

DOUGLASS, A. E.
  1928 Climatic cycles and tree-growth. Carn. Inst. Wash. Pub. 289.

ENGELBRECHT, H. H.
  1961 Manual for use of the IBM 650 calculator for computing potential evapotranspiration and the water balance. Publications in Climatology 14(3). Lab. of Climatology. Drexel Institute of Technology, Centerton, N.J.

FERGUSON, C. W.
  1964 Annual rings in big sagebrush. Univ. of Ariz. Press. 95 p.
  1968 Bristlecone pine: science and esthetics. Science 159: 839-846.
  In Press A 7104-year annual tree-ring chronology for bristlecone pine, *Pinus aristara,* from the White Mountains, California. Tree-ring Bull. 29(3-4).

FRITTS, H. C.
  1958 An analysis of radial growth of beech in a central Ohio forest during 1954-1955. Ecol. 39: 705-720.
  1962 An approach to dendroclimatology: screening by means of multiple regression techniques. Jour. Geophysical Res. 67: 1413-1420.
  1963 Computer programs for tree-ring research. Tree-Ring Bull. 25(3-4): 2-7.
  1965 Tree-ring evidence for climatic changes in western North America. Mo. Weather Rev. 93: 421-43.
  1966 Growth-rings of trees: their correlation with climate. Science 154: 973-79.

FRITTS, H. C., AND E. C. FRITTS.
  1955 A new dendrograph for recording radial changes of a tree. For. Sci. 1: 271-76.

FRITTS, H. C., AND N. HOLOWAYCHUK.
  1959 Some soil factors affecting the distribution of beech in a central Ohio forest. Ohio Jour. Sci. 59: 167-86.

FRITTS, H. C., D. G. SMITH, AND M. A. STOKES.
  1965a The biological model for paleoclimatic interpretation of Mesa Verde tree-ring series. Amer. Antiquity 31 (2, part 2): 101-21.

FRITTS, H. C., D. G. SMITH, J. W. CARDIS, AND C.A. BUDELSKY.
  1965b Tree-ring characteristics along a vegetation gradient in northern Arizona. Ecol. 46: 393-401.

FRITTS, H. C., D. G. SMITH, C. A. BUDELSKY, AND J. W. CARDIS
  1965c The variability of ring characteristics within trees as shown by a reanalysis of four ponderosa pine. Tree-Ring Bull. 27(1-2): 3-18.

GATES, D. M.
  1965 Energy, plants, and ecology. Ecol. 46: 1-13.

GLOCK, W. S.
  1955 Tree growth II. Growth rings and climate. Botan. Rev. 21(1-3): 73-188.

GLOCK, W. S., R. A. STUDHALTER, AND S. R. AGERTER.
  1960 Classification and multiplicity of growth layers in the branches of trees. Smithsn. Inst. Misc. Collect. 140(1).

GLOCK, W. S., AND S. R. AGERTER.
  1963 Anomalous patterns in tree rings. Endeavour 22: 9-13

GLOCK, W. S., E. M. GAINES, AND S. R. AGERTER.
  1964 Soil moisture fluctuations under two

ponderosa pine stands in northern Arizona. U. S. Forest Service Research Paper RM-9.

GLOCK, W. S., AND S. R. AGERTER
1966 Tree growth as a meteorological indicator. Int. J. Biometeor. 10: 47-62.

HARPER, W. G.
1957 Morphology and genesis of calcisols. Soil Sci. Soc., Am. Proceedings 21: 420-24.

HORTON, J. S.
1955 Use of electrical soil-moisture units in mountain soils. Proc. 23rd Am. Meeting Western Snow Conf., Portland, Oregon. pp. 20-26.

JOHANSEN, D. A.
1940 Plant microtechnique. McGraw-Hill, New York. 523 p.

KOZLOWSKI, T. T., AND T. KELLER.
1966 Food relations of woody plants. Botan. Rev. 32: 293-382.

KRAMER, P. J., AND T. T. KOZLOWSKI.
1960 Physiology of trees. McGraw-Hill, New York. 642 p.

LAMARCHE, V. C.
1963 Origin and geological significance of buttress roots of bristlecone pines, White Mountains, California. U. S. Geol. Survey Prof. Paper 475-C: 148-49.
1967 Spheroidal weathering of thermally metamorphosed limestone and dolomite, White Mountains, California. Geol. Survey Research 1967. U. S. Geol. Survey Prof. Paper 575-C: 32-37.
1968 Rates of slope degradation as determined from botanical evidence, White Mountains, California. Geol. Survey Prof. Paper 352-1; 341-77.
1969 Environment in relation to age of bristlecone pines. Ecol. 50(1): 53-59.

LIBBY, W. F.
1963 Accuracy of radiocarbon dates. Science 140: 278-280.

MACDOUGAL, D. T.
1936 Studies in tree-growth by the dendrographic method. Carn. Inst. Wash. Pub. 462.

MATALAS, N. C.
1962 Statistical properties of tree-ring data. Pub. Int. Assoc. Sci. Hydrology 7: 39-47.

MITSCHERLICH, E. A.
1909 Des Gesetz des Minimums und das Gesetz des abnehmenden Bodenertrags. Landw. Jahrb. 38: 537-552.

MOONEY, H. A., G. ST. ANDRÉ, AND R. D. WRIGHT.
1962 Alpine and subalpine vegetation patterns in the White Mountains of California. Amer. Midl. Natur. 68: 257-273.

MOONEY, H. A., MARDA WEST, AND ROBERT BRAYTON
1966 Field measurements of the metabolic responses of bristlecone pine and big sagebrush in the White Mountains of California. Bot. Gaz. 127: 105-113.

ST. ANDRÉ, G., H. A. MOONEY, AND R. D. WRIGHT.
1965 The pinyon woodland zone in the White Mountains of California. Amer. Midl. Natur. 73: 225-239.

SCHULMAN, EDMUND.
1954 Longevity under adversity in conifers. Science 119: 396-399.
1956 Dendroclimatic changes in semiarid America. Univ. of Ariz. Press, Tucson. 142 p.
1958 Bristlecone pine, oldest known living thing. Nat. Geog. Mag. 113: 355-372.

SCHULZE, E. D., H. A. MOONEY, AND E. L. DUNN
1967 Wintertime photosynthesis of bristlecone pine (Pinus Aristata) in the White Mountains of California. Ecol. 48: 1044-1047.

SNEDECOR, G. W.
1956 Statistical methods applied to experiments in agriculture and biology. Iowa State Coll. Press, Ames.

SOIL SURVEY STAFF.
1951 Soil survey manual. U. S. Dept. Agric. Handbook No. 18. U. S. Govt. Printing Office, Washington. 503 p.

THORNTHWAITE, C. W., AND J. R. MATHER.
1955 The water balance. Publications in Climatology 8(1). Lab. of Climatology. Drexel Institute of Technology, Centerton, N.J.

VERNER, LEIF.
1961 A new type of dendrometer. Univ. of Idaho, Moscow (manuscript).

WRIGHT, R. D.
1963 Some ecological studies on bristlecone pine in the White Mountains of California. Ph.D. Thesis. Univ. of Calif., Los Angeles. 118 p.

WRIGHT, R. D., AND H. A. MOONEY.
1965 Substrate-oriented distribution of bristlecone pine in the White Mountains of California. Amer. Midl. Natur. 73: 257-284.